METHODS IN PLANT VIROLOGY

Methods in Plant Pathology Volume 1

Methods in Plant Virology

STEPHEN A. HILL
Agricultural Development and Advisory Service
Ministry of Agriculture, Fisheries and Food
Cambridge

Published on behalf of the
British Society for Plant Pathology by
Blackwell Scientific Publications
Oxford London Edinburgh
Boston Palo Alto Melbourne

Editorial offices:
Osney Mead, Oxford, OX2 0EL
8 John Street, London, WC1N 2ES
9 Forrest Road, Edinburgh, EH1 2QH
52 Beacon Street, Boston
Massachusetts 02108, USA
706 Cowper Street, Palo Alto
California 94301, USA
99 Barry Street, Carlton
Victoria 3053, Australia

First published 1984

set by Enset (Photosetting),
Midsomer Norton, Bath, Avon
and printed in
Great Britain by
the Alden Press, Oxford

DISTRIBUTORS

USA and Canada
Blackwell Scientific Publications Inc
PO Box 50009, Palo Alto
California 94303

Australia
Blackwell Scientific Publications Inc
31 Advantage Road, Highett
Victoria 3190

British Library
Cataloguing in Publication Data

Hill, Stephen A.
Methods in plant virology.—(Methods in plant pathology, ISSN 0266-5514; v.1)
1. Plant viruses
I. Title II. Series
581.2′34 QR351

ISBN 0-632-00995-0

Contents

Preface

The preparation of guides or schedules describing methodology is difficult. A preliminary consideration must be the scope of the exercise: at what level of technique should one aim, what basic expertise can be assumed and to what degree of sophistication should the exercise extend? In researching this volume I have found a great number of general texts which as part of the broader consideration of the science of virology, present reviews of techniques without describing the techniques themselves. Similarly, even where methodology is tackled, many authors either assume basic virological knowledge, or do not include recent developments. As one involved with virus diagnosis, and working from a background of general plant pathological experience I have concentrated in this volume on methods for virus diagnosis. Thus, many of the techniques required for plant virus research are omitted, these may form the basis for subsequent volumes in this series. In particular, electrophoretic techniques have not been described and virus purification and antiserum production are covered very superficially. New approaches to serodiagnosis may result from the development of monoclonal antisera which have none of the host protein antibodies associated with traditionally produced antisera. The very recent advances into virus detection using radioactively labelled complementary DNA probes have not yet found wide usage, but will undoubtedly do so for the viroid induced diseases, and perhaps for some true viruses. The apparent exponential rate of development of plant virological technique, which probably relates to the increasing comparability of techniques for animal virus study, probably means that this volume will soon be out of date. However, even the most ambitious student of plant virology must assimilate the basic techniques before attempting greater sophistication and hopefully this volume will provide the first rungs on the ladder.

The techniques are arranged in successive chapters in a sequence which to some extent may provide a route to virus characterization. It is impossible, however, to define pathways strictly since, for example, in some circumstances quick electron microscopy may provide valuable first steps in an investigation or basic serology may be appropriate.

Wherever possible the techniques are described first in outline to provide a basic understanding, the materials required are listed and then the step-by-step procedure. Factors affecting the success of the test and interpretation of the results etc., are provided and where appropriate useful references are listed. Detailed recipes for component 'ingredients' etc., are appended as necessary at the end of each chapter.

As will become evident, the different techniques each contribute some knowledge of the virus identity, but also of its behaviour and characteristics. Relationships between viruses may be investigated; using some others simply provides sensitive rapid detection for large numbers of samples. Absolute virus characterization demands systematic evaluation in a range of tests. Serological tests relate unknown viruses to those previously described. Electron microscope serology has the additional advantage of providing the reassurance of actually seeing virus particles. None of the tests described is able to prove absence of virus conclusively.

Finally, the objective of this volume has been to provide a basis for the investigation of plant viruses. As far as possible only basic biological and chemical knowledge is assumed. It should serve as a bench book but still, I hope, provide a contribution to the understanding of plant viruses.

Acknowledgments

I should like to thank my colleagues for their valuable help and valued advice in the preparation of this book, Dr Preece for his part in stimulating it, and my family for their help and forbearance during its production.

1

Introduction

For the aspiring plant virologist, or for those whose interest in the science is secondary, the search for simple methodology can be difficult. Whilst many of the techniques still essential in the study of plant viruses are long established, others are very recently described. The former are considered in principle by the more general texts with little or no practical guidance, and the latter appear mostly in the virological journals. It is often necessary for students, or those new to virology to spend some time working through methods in scientific papers, abbreviated for publication and lacking in essential simple details. Often in scientific literature the author assumes that the reader is a fellow virologist, familiar with the basic principles, and there may often be cross referencing to related work. Thus, to find the practical detail may involve a search of several papers. This volume attempts to present basic methods in plant virology and should serve as a bench book.

However, no volume of 'methods' can ever be complete since new methods are frequently described. Similarly, it would be impossible to present all the minor modifications to the methods described by different practitioners. Such variations do not usually make fundamental changes and each laboratory may have its own peculiar local problems which justify alterations. Viruses are of course biological entities and therefore subject to change, and methods may need some adaptation to each laboratory circumstance.

The methods chosen for description are mostly those in common use for virus diagnosis, but most have other uses. Indeed there is perhaps a need for a clear definition; 'testing for virus' can simply mean testing to demonstrate the presence of virus in host material. Further 'testing' may characterize the virus, indicate the quantity of virus, illustrate its relationship with other viruses, vectors, etc. However, for many purposes the need to confirm (a) virus presence and (b) its identity, are fundamental to any investigation.

Where possible, the methods described are accessible even to the casual plant virologist, without the need for sophisticated equipment. Electron microscopy may be the exception, although access to transmission electron microscopes is much easier nowadays. Perhaps in some circumstances, the requirement for antisera for serodiagnosis or immune electron microscopy may be equally limiting. This volume does not describe methods for virus purification and antiserum production. Antisera to a limited range of plant viruses can be purchased (see p. 124) and this may present a solution where a research programme is being started. However, antiserum production is costly and its misuse or waste is to be discouraged. In many tests the amount of serum required is very little and such quantities may often be 'borrowed' or exchanged.

Preliminary diagnosis of plant viruses may sometimes be achieved using one technique alone. Such is the case with sap transmission testing. However, no dichotomous key for such diagnosis exists, the information being scattered through publications each often concerning individual virus/host combinations. Within this volume (see pp. 42–61) are two approaches to diagnostic routes using sap transmission tests. These have been constructed over many years and owe a great deal to the elder statesmen of plant virology. Diagnosis is usually achieved by a combination of techniques. Histological tests which are comparatively quick and require little sophisticated equipment may provide preliminary information. Electron microscopy of crude sap extracts may also provide a comparatively simple starting point. These, together with the symptom characteristics, host range etc., may be sufficient to confirm virus presence and begin virus characterization which can then be confirmed by more extensive serology and sap transmission tests. However, even where all the techniques are available no diagnostic guide exists.

Faced with the apparently infinite range of plant viruses whose vernacular names provide no guide to generic relationships, the aspiring diagnostician may lose heart. However, the virus groups defined by the International Committee on Taxonomy of Viruses [1] provide a rational basis from which a diagnostic route may be chosen. The virus groups have been defined by establishing the common properties of their members, and some of these features may be valuable in diagnosis. This may be illustrated by comparing some features of members of the tobamovirus group with those of the luteoviruses (see pp. 6, 8).

The tobamoviruses may produce crystalline inclusion bodies in cells, the luteoviruses may cause necrosis or blocking of vascular tissue. Tobamoviruses invade their hosts systemically reaching high concentration, and are easily seen in electron microscope preparations of crude sap. In contrast, the luteoviruses are phloem-restricted and cannot be detected in simple electron microscope preparations. Tobamoviruses cause mosaics or mottles and are easily sap transmissible, whilst luteoviruses cause leaf yellowing or reddening and can only be transmitted using their natural aphid vectors. These, together with easily determined physical properties, e.g. thermal inactivation point, longevity in sap, are important indicators of the nature of viruses, and provide a guide to the best means of diagnosis. The characteristics of the well established virus groups and lists of their members are provided in the appendix to this chapter (see p. 4).

Properties of some of the virus groups and many of the individual viruses are described in the CMI/AAB descriptions series [2]. The need for basic technique in virus detection in plants is more acute nowadays than ever before. Apart from the effects of intensification of cropping and change in agronomic practice on virus and vector incidence, the damaging potential of viruses is more widely appreciated. Particularly in the amenity plant industry, the trend to micropropagation has been marked and the need for virus-free starting material and for health monitoring has been quickly recognized. Thus, the practice of virus diagnosis, traditionally confined to research and advisory and statutory extension services has now become a commercial requirement.

A note of warning is appropriate before the techniques described in the following chapters are employed. Whichever test is used it is only ever possible to demonstrate presence of virus, never complete absence. Exhaustive testing and use of the more sensitive techniques reduces the likelihood of virus presence being overlooked but cannot guarantee it. Further, particularly with serological testing it is essential to undertake parallel control experiments since reactions may not be virus-induced.

The techniques described represent the very basic of their kind and are intended as a primer for more sophisticated steps later. Many can be done using crude sap preparations and using comparatively simple equipment. Provided wherever possible the execution of a particular test is paralleled by an understanding of the underlying principles then successful diagnosis should be possible.

References

1. Matthews, R.E.F. (1979) Classification and Nomenclature of Viruses. Third Report of the International Committee on Taxonomy of Viruses. *Intervirology* **12**, (3–5).
2. *CMI/AAB Descriptions of Plant Viruses.* Commonwealth Agricultural Bureaux, Farnham Royal, Bucks.

Appendix: Virus Groups

As more information about the character of viruses becomes available it is possible to group certain viruses with similar characteristics. A knowledge of the relationships between viruses and the group to which they belong can help in predicting virus behaviour. The groups of viruses so far defined and some of their more commonly encountered members are presented below. Complete lists for some groups can be found in the CMI/AAB group descriptions. (Figures in brackets are CMI/AAB descriptions numbers.)

Viruses with rod-shaped or filamentous particles

Carlaviruses (*car*nation *la*tent *virus* group) (259)
Flexuous filamentous rods 620–700 nm long.
Contain single-stranded RNA.
Thermal inactivation point (TIP) 55–70°C (10 min).
Survive in sap a few days at 20°C.
Concentration in sap 20–100 mg l^{-1}.
Symptoms: often slight or none.
Narrow host range.
Most aphid transmitted, non-persistent.
Examples: alfalfa latent (211); carnation latent (61); chrysanthemum B (110); cowpea mild mottle (140); hop mosaic (241); lily symptomless (96); narcissus latent (170); pea streak (112); poplar mosaic (75); potato M (87); potato S (60); red clover vein mosaic (22); shallot latent (250).

Closteroviruses (from Greek *Kloster,* a thread; very long thin particles)
Very long flexuous rods 1250–2000 nm long.
Contain single-stranded RNA.
TIP 45–55°C, survive in sap a few days at 20°C.
Concentration in sap 40–100 mg l^{-1}.
Moderately wide host range.

Symptoms: variable but may affect older leaves with yellowing or necrotic spots.

Sap transmissible with difficulty, aphid transmission of some, semi-persistent.

Examples: apple chlorotic leaf spot (30); beet yellows (13); beet yellow stunt (207); carnation necrotic fleck (136); citrus tristeza (33); heracleum latent (228); lilac chlorotic leafspot (202); wheat yellow leaf (157).

Hordeiviruses (barley stripe mosaic virus group)

Straight tubular rods 110–160×20–25 nm.

Contain single-stranded RNA.

TIP 63–70°C, survive in sap several days or weeks.

Sap transmissible but rather restricted host range.

Some members seed or pollen-borne.

No known insect vectors.

Example: barley stripe mosaic (68).

*Potexviruses (pot*ato *X virus* group) (200)

Flexuous filamentous rods 470–580 nm long.

Contain single-stranded RNA.

TIP 65–80°C, survival in sap several months at 20°C.

Concentration in sap, up to 500 mg l^{-1}.

Symptoms: mosaics, mottles or ringspots.

Restricted natural host range, highly contagious.

Readily spread by plant contact.

Easily sap transmissible.

No natural insect vector.

Examples: cactus X (58); cassava common mosaic (90); clover yellow mosaic (111); cymbidium mosaic (27); daphne X (195); hydrangea ringspot (114); narcissus mosaic (45); papaya mosaic (56); potato aucuba mosaic (98); potato X (4); viola mottle (247); white clover mosaic (41).

Potyviruses (*pot*ato *Y virus* group) (245)

Flexuous filamentous rods 720–900 nm long.

Contain single-stranded RNA.

TIP 50–60°C, survive in sap a few days at 20°C.

Concentration in sap 5–35 mg l^{-1}.

Symptoms: usually mosaics and mottles.

Some members seed-borne.

Easily sap transmissible but some members have rather restricted host range.

Aphid transmitted, non-persistent.

Examples: bean common mosaic (73); bean yellow mosaic (40); bearded iris mosaic (147); beet mosaic (53); bidens mottle (161); carnation vein mottle (78); carrot thin leaf (218); celery mosaic (50); clover yellow vein (131); cocksfoot streak (59); cowpea aphid-borne mosaic (134); dasheen mosaic (191); guinea grass mosaic (190); henbane mosaic (95); hippeastrum mosaic (117); iris mild mosaic (116); leek yellow stripe (240); lettuce mosaic (9); narcissus yellow stripe (76); onion yellow dwarf (158); papaya ringspot (84); parsnip mosaic (91); passionfruit woodiness (122); pea seed-borne mosaic (146); peanut mottle (141); pepper mottle (253); pepper veinal mottle (104); peru tomato (225); plum pox (70); pokeweed mosaic (97); potato A (54); potato Y (37, 242); soybean mosaic (93); sugarcane mosaic (88); tobacco etch (55, 258); tulip breaking (71); turnip mosaic (8); watermelon mosaic (63).

Tobamoviruses (*toba*cco *mo*saic *virus* group) (184)

Straight tubular rods 300×18 nm.

Contain single-stranded RNA.

TIP more than 90°C, survive in sap months or years.

Concentration in sap more than 1 mg ml^{-1}.

Symptoms: mosaics and mottles.

Highly contagious, readily sap transmitted.

No insect vector, spread by foliage contact and in soil with root debris.

Most members have wide host range.

Examples: beet necrotic yellow vein (144); broad bean necrosis (223); cucumber green mottle mosaic (154); frangipani mosaic (196); nicotiana velutina mosaic (189); odontoglossum ringspot (155); peanut clump (235); potato mop-top (138); ribgrass mosaic (152); sunn-hemp mosaic (153); tobacco mosaic (type strain) (151); tomato mosaic (156); wheat (soil-borne) mosaic (77).

Tobraviruses (*tob*acco *ra*ttle *virus* group)

Straight tubular particles of two lengths:

(i) Longer particle 180–210 nm according to strain, infectious on their own;

(ii) Shorter particles characteristic of the strain, not infectious on their own (code for coat protein).

Bipartite genome—both parts needed to produce complete virions with protein coat.

Contain single-stranded RNA.
TIP 70–80°C, survive in sap several months.
Concentration in sap 20–100 mg l^{-1}.
Symptoms: mostly necrotic blotches and streaks.
Wide host range, readily sap transmitted.
Spread by soil nematodes, virus persists several weeks in vector (Trichodorus+Paratrichodorus).
Examples: pea early-browning (120); tobacco rattle (12).

Viruses with isometric particles

Bromoviruses (*bro*me *mo*saic *virus* group) (215)
Isometric particles 26 nm diameter.
Contain single-stranded RNA, of three distinct types.
TIP 70–95°C according to virus, survival in sap very variable.
Concentration in sap 0·5–5 mg l^{-1}.
Sap transmissible, some have beetle vectors.
Symptoms: mottle, restricted host range.
Examples: broad bean mottle (101); brome mosaic (3, 180); cowpea chlorotic mottle (49); melandrium yellow fleck (236).

Caulimoviruses (*cauli*flower *mo*saic *virus* group)
Isometric particles 50 nm diameter.
Contain double-stranded DNA.
TIP 75–80°C, survive in sap a few days.
Concentration in sap less than 10 mg l^{-1}.
Symptoms: mosaic, mottle, vein banding or rings.
Sap transmissible but restricted host range.
Aphid transmitted, semi-persistent.
Examples: carnation etched ring (83); cauliflower mosaic (24, 243); dahlia mosaic (51); strawberry vein banding (219).

Comoviruses (*co*wpea *mo*saic *virus* group) (199)
Isometric particles 28 nm diameter.
Two particle types contain single-stranded RNA, a third is empty.
TIP 60–80°C, survive in sap one to several weeks.
Concentration in sap 50–200 mg l^{-1}.
Symptoms: mosaic and mottle, narrow host ranges.
Sap transmissible.
Beetle vector, virus persists several days in beetle.
Examples: andean potato mottle (203); bean pod mottle (108); bean rugose mosaic (246); broad bean stain (29); cowpea mosaic (47,

197); cowpea severe mosaic (209); echtes Ackerbohnenmosaik (20); quail pea mosaic (238); radish mosaic (121); red clover mottle (74); squash mosaic (43).

Cucumoviruses (*cucu*mber *mo*saic *virus* group)

Isometric particle 28 nm diameter.

Three particle types contain single-stranded RNA of different molecular weights.

TIP 60–70°C, survive in sap a few days.

Concentration in sap 20–200 mg l^{-1}.

Cause transmissible wide host range.

Aphid transmitted, non-persistent. Seed-borne in a few hosts.

Examples: cucumber mosaic (1, 213); peanut stunt (92); robinia mosaic (65); tomato aspermy (79).

Geminiviruses (maize streak virus group—from Latin *gemini*—characteristic double particle)

Isometric particles 18–20 nm occurring in pairs.

Each particle containing single-stranded DNA.

TIP 60°C, survive more than one day in sap.

Dilution end-point not less than 10^{-3}.

Symptoms: streak and mosaics.

Narrow host range—may be sap transmissible.

Persistently transmitted by whiteflies or leafhoppers.

Examples: bean golden mosaic (192); beet curly top (210); chloris striate mosaic (221); maize streak (133); tobacco leaf curl (232).

Ilarviruses (*i*sometric *la*bile-*r*ingspot *virus* group)

Three types of particle, spherical 26–35 nm.

Divided genome in four parts all needed for infectivity.

Contain single-stranded RNA.

TIP 50–60°C, survive in sap 2–10 days at 20°C.

Symptoms: mottle streak and line pattern.

Sap transmissible, wide host range.

Some are pollen-borne in flower bearing plants.

Examples: apple mosaic (83); black raspberry latent (106); citrus leaf rugose (164); elm mottle (139); lilac ring mottle (201); prune dwarf (19); prunus necrotic ringspot (5); tobacco streak (44); tulare apple mosaic (42).

Luteoviruses (barley yellow dwarf virus group (*luteo* = yellow))

Isometric particles 25 nm diameter.

Contain single-stranded RNA.

TIP 65–70°C, concentration usually less than 10 mg l^{-1}.

NOT sap transmissible.
Aphid vectors, usually highly specific, persistent.
Symptoms: yellowing, reddening of plants and dwarfing.
Examples: barley yellow dwarf (32); beet western yellows (89); carrot red leaf (249); potato leafroll (36); soy-bean dwarf (179); tobacco necrotic dwarf (234).

Nepoviruses (*ne*matode transmitted *po*lyhedral *viruses*) (185)
Isometric particles 28 nm diameter.
Three particle types; two with different RNA, one empty.
Divided genome—both RNA types necessary for infection.
Contain single-stranded RNA.
TIP 55–70°C, survive in sap a few days or weeks.
Concentration in sap often 10 mg l^{-1}.
Symptoms: ringspot/mottle.
Wide host range.
Sap transmissible.
Nematode-borne, virus persists weeks in vector.
Seed-borne.
Examples: arabis mosaic (16); arracacha A (216); artichoke Italian latent (176); cacao necrosis (173); cherry leaf roll (80); chicory yellow mottle (132); grapevine Bulgarian latent (186); grapevine chrome mosaic (103); grapevine fanleaf (28); hibiscus latent ringspot (233); lucerne Australian latent (225); mulberry ringspot (142); myrobalan latent ringspot (160); peach rosette mosaic (150); potato black ringspot (206); raspberry ringspot (6, 198); strawberry latent ringspot (126); tobacco ringspot (17); tomato black ring (38); tomato ringspot (18).

Pea Enation Mosaic Virus Group (monotypic)
Isometric particles 30 nm diameter.
Two types of particles both needed for infectivity.
Contain single-stranded RNA.
TIP 55–60°C, survive in sap a few days.
Concentration in sap 5–15 mg l^{-1}.
Symptoms: mottle and enation.
Narrow host range, sap transmissible.
Aphid-borne, semi-persistent (last weeks in vector).
Example: pea enation mosaic (25, 257).

Plant Reoviruses
I Phytoreoviruses
Isometric particles *c.* 70 nm diameter.

Contain double-stranded RNA.
TIP 45–50°C, survive in sap 2 days at 4°C.
Dilution end-point 10^{-4}–10^{-5}.
Symptom: stunting effect.
Narrow host range.
Persistently transmitted by leafhoppers.
Virus multiplies in vector and passes through vector egg stage.
Examples: rice dwarf (102); wound tumour (34).

II Fijiviruses
Polyhedral particles 65–70 nm.
Contain double-stranded RNA.
Unstable.
Symptoms: dwarfing and sterility.
Narrow host range.
Persistently transmitted by leafhoppers.
Virus multiplies in vector and passes through vector egg stage.
Examples: maize rough dwarf (72); oat sterile dwarf (217); pangola stunt (175); rice black-streaked dwarf (135); sugarcane Fiji disease (119).

Tobacco Necrosis Virus Group
Isometric particles 26 nm diameter.
Contain single-stranded RNA.
TIP 65–95°C, survive several months.
Easily sap transmissible with wide host range, (mostly local lesion reactions).
Vector, the soil fungus *Olpidium*—zoospores retain fungus for a few hours.
Example: tobacco necrosis (14).

Tomato Spotted Wilt Virus Group (monotypic)
Isometric particles 70–90 nm diameter.
Contains single-stranded RNA.
TIP 42°C, survive in sap for a few hours at 20°C, concentration in sap less than 20 mg l^{-1}.
Symptoms: ringspot and necrosis.
Wide host range.
Sap transmissible (with a reducing agent).
Thrip transmitted—persists in vector for weeks.
Example: tomato spotted wilt (39).

Tombusviruses (*tom*ato *bus*hy stunt *virus* group)
Isometric particles 30 nm diameter.

Contain single-stranded RNA.
TIP 85–90°C, survive in sap a few weeks at 20°C.
Concentration in sap often 20–200 mg l^{-1}.
Wide host range: sap transmissible, no insect vector known; some members soil-borne but no soil vector known; physical contamination.
Examples: tomato bushy stunt (69); possible members: carnation mottle (7); cymbidium ringspot (178); galinsoga mosaic (252); narcissus tip necrosis (166); pelargonium flower-break (130); saguaro cactus (148); tephrosia symptomless (256); turnip crinkle (109).

Tymoviruses (*t*urnip *y*ellow *mo*saic *virus* group) (214)
Isometric particles 29 nm diameter.
Contain single-stranded RNA.
TIP 70–90°C, survive in sap a few weeks at 20°C.
Concentration in sap 50–500 mg l^{-1}.
Symptoms: mosaic and mottle, narrow specialized host ranges. Readily sap transmissible.
Beetle vectors remain infective several days.
Examples: belladonna mottle (52); cacao yellow mosaic (11); clitoria yellow vein (171); desmodium yellow mottle (168); eggplant mosaic (124); erysmium latent (222); kennedya yellow mosaic (193); okra mosaic (128); scrophularia mottle (113); turnip yellow mosaic (2, 230); wild cucumber mosaic (105).

Viruses with bacilliform particles

Alfalfa Mosaic Virus Group (= lucerne mosaic)
Monotypic. Bacilliform particles of at least four kinds 18×58, 18×48, 18×36 and one near spherical at 18 nm diameter.
Divided genome. Three largest particles needed for infectivity.
Contain single-stranded RNA.
TIP 60–70°C, survive in sap for a few days.
Concentration in sap 20–500 mg l^{-1}.
Symptoms: mottle mosaic and ringspot.
Wide host range, sap transmissible.
Aphid-borne, non-persistent (seed-borne in some hosts).
Example: alfalfa mosaic virus (46, 229).

Plant Rhabdoviruses (244)
Bacilliform particles 200–350×70–95 nm.

Contain single-stranded RNA.
TIP 50–52°C, readily becomes inactivated in sap.
Concentration in sap is 1–10 mg l^{-1}.
Yellows often becoming necrotic.
Some members sap transmissible.
Transmitted persistently by leaf-sucking arthropods (aphids, and leafhoppers usually).
Some may multiply in their insect vectors.
Sub-group I: broccoli necrotic yellows (85); lettuce necrotic yellows (26); raspberry vein chlorosis (174); strawberry crinkle (163); wheat (American) striate mosaic (99).
Sub-group II: cereal chlorotic mottle (251); eggplant mottled dwarf (115); maize mosaic (94); orchid fleck (183); potato yellow dwarf (35); rice transitory yellowing (100); sonchus yellow net (205); sowthistle yellow vein (62).

2

Histological and Other Basic Methods

Apart from the obvious expression of symptoms caused by virus infection, characteristic cytological and histological changes may occur. With some host/virus combinations these may be valuable in the preliminary stages of virus diagnosis using staining techniques and conventional light microscopy. For the most part, observation of cytological changes has been confined to research characterization of viruses and has found limited routine application. Observations of structures such as pinwheel inclusions, often found in plants infected with potyviruses, may provide useful supportive evidence in an investigation by electron microscope.

Inclusion Bodies

The inclusion bodies sometimes found in plant tissue infected with certain viruses are readily observed by light microscopy. These structures, most often in epidermal cells of leaves and stems, were mistaken for living organisms by early researchers. Further investigation has brought a clearer understanding of the formation, composition and diversity of the inclusion bodies associated with different viruses. Whilst there is still much to be learned concerning the significance of inclusion bodies, with a limited range of virus infections their observations can be a useful first step in diagnosis. Several authors have classified inclusion bodies into a number of groups and to some extent these relate to the causal viruses. Some viruses which produce inclusion bodies are: bean yellow mosaic; turnip mosaic; cauliflower mosaic; dahlia mosaic; red clover vein mosaic; tobacco etch; tobacco mosaic; watermelon mosaic.

Epidermal strips are often suitable for observation of inclusion bodies, but in some hosts sections must be cut where epidermal strips cannot be prepared, or when inclusion bodies are located in specific tissues and are absent from the epidermis. Trichomes from stems or hairy leaves of infected plants are often good for observation and may not need fixation or staining.

Preparation of material for light microscope examination may simply involve staining or even simple observation in tap water. Rapid staining procedures involve use of stains such as trypan blue, (0·5% trypan blue in 0·9% NaCl) in which nuclei turn blue quickly and inclusions stain strongly. For some inclusion bodies aqueous phloxine (1%) may be better, inclusion bodies staining bright red. A combination of trypan blue and phloxine stains has also been described. A variety of other staining procedures are described, differing in complexity [2].

Fixation of tissue may be advantageous in some studies. Fixed cells are killed but the internal structure is retained and entry of stains facilitated. A variety of procedures for fixation and embedding, staining and permanent preparation are used [2].

Histological Changes

The observation of histological changes, specifically those which resulted from infection by luteoviruses, has been widely used in diagnosis. To a large extent such tests were developed when sero-diagnosis of these low concentration viruses was impossible and there was a dearth of alternative rapid test techniques. Thus, in the absence of critical competition, the fact that histological tests tend to be somewhat subjective could be overlooked; they may not be so valuable nowadays. However, in circumstances where sophisticated equipment for IEM or serodiagnosis is not available histological tests may still be used; e.g. the following two histological phenomena induced by infection of potato by potato leaf roll virus.

Phloem necrosis

Infection by potato leaf roll virus leads to necrosis of phloem tissue in potato stems, and also in mature tubers. Sectioning stems, staining and observation by light microscopy can provide a diagnosis of this virus.

Procedure

1 Cut a piece of main stem of potato suspected of being infected. This should be cut from 1 inch below soil level to the sixth node. Trim off leaves and side shoots.

2 Cut thin transverse sections *through nodes.*
3 Stain in phloroglucinol (1% in 50% alcohol) on a slide for 1 minute, then drain.
4 Immerse sections in a few drops of 50% HCl for 1 minute then drain.
5 Mount in water.
6 Examine by conventional light microscopy at 100×.

In healthy plant sections, the xylem tissue should stain purplish red and the phloem remain colourless. Fibres may be pink. If leaf roll virus is present, some primary phloem strands stain yellowish red, and often the cambium is irregular with uneven development of secondary xylem.

The Igel Lange test for callose deposition

Callose is deposited in phloem sieve tubes in abnormal quantities as a response to injury, toxins or infection with phloem-restricted viruses. Callose is formed to seal sieve tubes which have ceased to function, sometimes temporarily, during winter and may be rapidly dissolved in spring. Callose deposition increases as the plant ages. The Igel Lange test has been widely used to detect potato leaf roll virus infection in seed potato tubers, and in stems of infected plants. The stain used for callose detection is 1% aqueous resorcin blue (see Appendix p. 17) which turns callose deep blue.

To detect callose deposition in stems

Select stems of moderate thickness and prepare fresh, relatively thick longitudinal slices 1–2 mm thick. Stain for 10 minutes in resorcin blue (see Appendix p. 17), then examine using a binocular microscope at a magnification of 25×. Callose in stems may be obscured by chlorophyll and the test is more effective for tubers.

Detection of callose in potato tubers [1]

Procedure

1 Wash mature tubers to be tested to remove excessive soil. (Best results are from tubers harvested for more than 4 weeks, which have been stored at 16°C and are unsprouted.)
2 Cut longitudinal slices 2–3 mm thick and 3–4 mm wide from

the heel end (stolon end) of the tuber. This may be done using a razor, but where the test has been used routinely a double-bladed knife has been constructed.

3 Stain sections for 10 minutes in resorcin blue (see Appendix p. 17).

4 Examine at 25× magnification under a binocular microscope.

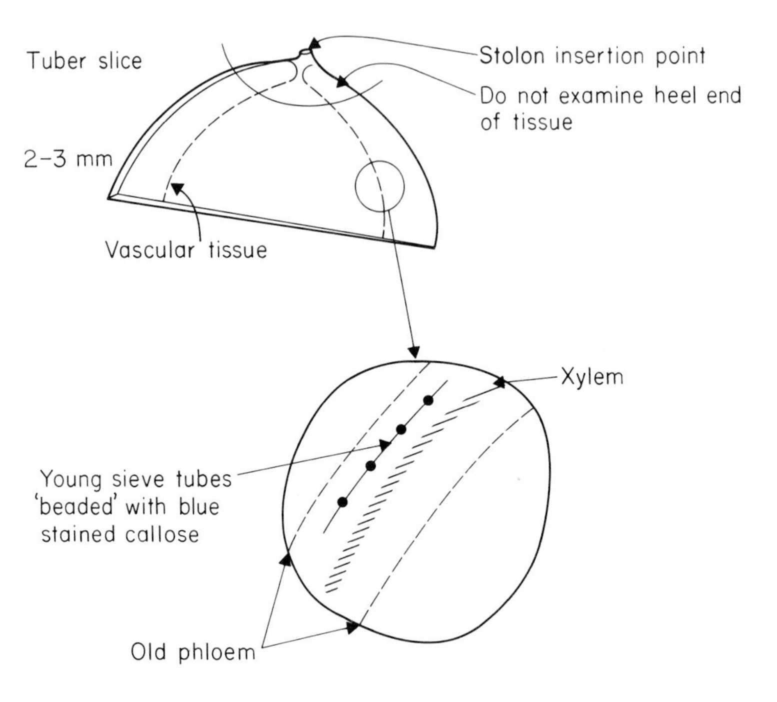

Fig. 2.1. Igel Lange test for callose deposition by potato leaf roll virus infection.

Old sieve tubes whether diseased or not often contain a lot of callose and may be misleading. Younger phloem tissue near the cambial layer should be examined (Fig. 2.1). The extreme heel end, the stolon insertion point, may also contain much callose and should be ignored. Interpretation of acceptable normal callose levels as distinct from virus-induced callose must be determined by experience. Varietal differences in normal callose deposition have to be taken into account, so that unless the test is done regularly it may be difficult to interpret objectively. Tubers infected during the growing season (primary infection) may have more callose than those produced by infected plants (secondary infection).

References

1. Bokx, J.A. de (1967) The callose test for the detection of leafroll virus in potato tubers. *European Potato Journal,* **10,** 221–234.
2. Rubio-Huertos, M. (1972) Inclusion bodies. In *Principles and Techniques in Plant Virology* (eds Kado, C.I. & Agrawal H.O.), p. 62. Van Nostrand Reinhold, New York.

Appendix: Igel Lange Test

Stock solutions of resorcin blue

Dissolve 10 g pure white resorcin (metadioxybenzol) in 1 litre distilled water and add 12 ml 25% ammonia (NH_4OH). Keep the mixture in an open container at room temperature for 10–14 days until the smell of ammonia disappears. The greenish blue stain may then be stored satisfactorily for up to a year but should be discarded as it darkens since it will then stain xylem and starch as well as callose.

3

Basic Virus Characterization and Storage

In preliminary consideration of which technique or combination of techniques should be used to study a particular virus, certain basic characteristics are often of value. Measures of thermal inactivation point, longevity *in vitro* and dilution end-point not only provide guidance as to how easily a particular virus may be investigated, but may also help in initial diagnosis if reference to the characteristics of the virus groups is made (see Chapter 1). A further useful measure in preliminary investigations is that which indicates the concentration of virus in sap extracts, the local lesion assay. For determination of all these, crude extracts of virus in host sap are used. Thus, the criteria established may vary according to a number of factors, e.g. age of source material, presence of virus inhibitors in sap, etc. However, provided it is accepted that the determinations provide only approximate data from comparative purposes only, they may still be of value.

Evaluation of thermal inactivation point, longevity *in vitro* and dilution end-point, as well as local lesion assay can only be achieved simply with viruses which are mechanically transmissible (see Chapter 4) since retention of infectivity must be measured. For those viruses for which the natural vector must be used to establish retention of activity, characterization is still possible but is more difficult and not so frequently attempted. More complex serological means to measure dilution end-point may be found in Chapter 5.

Preparation of crude sap extract of virus

Source material should consist of leaves from relatively young plants showing typical symptoms. Usually systemically infected leaves showing most distinct symptoms about 10–14 days after inoculation should be used. Leaves are chopped into smaller pieces, ground in a pestle and mortar and the resultant pulp squeezed through a muslin cloth to extract the sap. Attempts to standardize determination of basic virus characteristics have suggested the inclusion of buffers in crude sap extracts for test. However, most often no modification is made, except perhaps for the addition of reducing agents or inhibitors of oxidizing enzymes (see p. 31) to saps containing viruses known to be particularly unstable. Low-speed centrifugation to remove intact cells is sometimes recommended.

Thermal Inactivation Point

As infectious crude sap is heated the rate of virus inactivation depends not only on temperature but also on virus concentration, pH, the presence of oxidizing enzymes, etc. Thus, some variation due to the propagation host selected might be expected. The most commonly accepted definition of the thermal inactivation point of a virus in crude juice is the temperature required for complete inactivation following a 10-minute exposure. Usually the two temperatures between which complete inactivation occurred are quoted.

Procedure

1 Prepare 100 ml crude extract of sap containing virus from the source plants as described above.

2 Pipette 2 ml volumes of sap into each of ten thin-walled test tubes, taking care to avoid contamination of the tube walls with sap.

3 Heat a water bath (water of depth sufficient to immerse the tubes to 3 cm above the level in each tube) to the lowest required temperature within the range (45°C).

4 Immerse the first tube in the water bath standing a thermometer alongside (but not inside) the tube, and keep the temperature steady for 10 minutes.

5 Remove the tube after 10 minutes, cool in running water, label with temperature used and set aside for assay.

6 Repeat using successive tubes, raising the temperature by 5°C between each.

7 When all the tubes have been treated, inoculate separate test plants with the sap as described in Chapter 4 (an identical amount of abrasive should be added to each before inoculation).

8 Note the temperature (or temperature pair) at which total inactivation occurs.

N.B. It is wise to retain a small amount of the original extract for inoculation as a control. More precise determination of thermal inactivation point may be made by repeating the above procedure using a narrower range of 2°C steps.

Longevity *In Vitro*

Different viruses behave differently when allowed to age *in vitro* at room temperature. Loss of activity during such ageing may result from thermal inactivation or it may result from the action of oxidative enzymes, etc. Thus, the longevity *in vitro* illustrates quite different properties of the virus from those determining the thermal inactivation point. As with thermal inactivation point the results of longevity tests may be affected by a variety of factors, e.g. the source plant type, the room temperature etc. and for greatest accuracy reports should be accompanied by details of procedure. Longevity *in vitro* is defined as the time (hours, days or weeks) for which crude juice kept at room temperature remains infective. Unless some approximation of the longevity is already known it is usual to make initial tests at a series of intervals at a geometric progression (i.e. 1, 2, 4, 8, 16, 32 . . . days) until infectivity is lost. Once a preliminary value is determined tests may be repeated over a narrower range of shorter intervals.

Procedure

1 Prepare a crude extract of infected sap as before (see p. 18): 2 ml of sap will be needed for each time interval to be tested.

2 Pipette 2 ml aliquots of sap into a sufficient number of tubes to give the chosen range of time intervals (e.g. 0, 0·25, 1, 2, 4, 7, 10, 14 days = 8 tubes) and seal the tops of the tubes with a stopper or with 'Parafilm'.

3 Leave tubes for the prescribed intervals making sure that they are not exposed to direct sunlight or other variable heat or cooling influences.

4 After the prescribed interval each tube should be opened and the sap used (with abrasive added) to inoculate appropriate indicator plants (see Chapter 4).

5 Assess virus reactions in inoculated plants after an appropriate time and record the sap exposure time interval after which no infection occurs.

Dilution End-Point

The extent to which a virus multiplies or accumulates in its host is characteristic. Determination of the dilution of crude sap extracts at which infectivity is lost is therefore a useful pointer in virus characterization. As with the previous two criteria measured, dilution end-point values are affected by a great many factors—source plant, time since inoculation, etc.—so as before in reporting results it is best to provide some indication of the procedure and materials used. The dilution end-point is usually reported as being between two dilutions, i.e. between the highest dilution that was still infective and the next highest one. Dilutions on a logarithmic scale (1/1, 1/10, 1/100, 1/1000 . . . 1/10,000,000) are usually prepared in initial determinations with viruses of unknown character. A second test may then be prepared with closer dilutions when a more accurate end-point is determined.

Procedure

1 Prepare a crude sap extract as before.

2 Prepare a range of dilutions as follows: into the first of eight tubes pipette 2 ml of extract. Into the remaining seven tubes pipette 1·8 ml water using a clean pipette. From tube one (crude extract) extract 200 μl sap and transfer to tube two. Mix to give a 1/10 dilution. From tube two (1/10) transfer 200 μl to tube three. Mix to give a 1/100 dilution. Repeat with tubes four to eight, so that the greatest dilution is 1/10,000,000.

3 Immediately make separate inoculations from each tube dilution (adding an appropriate amount of abrasive) to several test plants (preferably 5) (see Chapter 4).

4 After an appropriate interval record symptoms and note the dilutions between which infectivity is lost.

In repeat tests, halving dilutions around that determined above should give a more exact value.

Local Lesion Assay

For mechanically transmissible viruses which produce local lesions, i.e. discrete lesions only on the leaf inoculated (see p. 26), the local lesion assay provides a means of measuring virus concentration. The number of local lesions has been shown to be inversely correlated with the dilution of the inoculum. Thus, it is possible to determine, for example, the rate at which virus infectivity is lost during thermal inactivation or longevity determinations. Often the rate of loss of infectivity is not constant throughout such tests.

Because the results of local lesion assays may be seriously affected by a variety of factors other than the concentration of active virus, experiments must be carefully designed. Designs must be such as to eliminate variation other than that induced by the changes in virus concentration. Early experiments showed that lesion numbers were very variable from the same inoculum when applied to different leaves on the same plant, and to different leaves from different plants. However, lesion numbers on opposite halves of the same leaf separately inoculated with the same dilutions are usually very close. Similarly, opposite leaves on the same branch or main stem usually react with equal numbers of lesions. Thus, halves of opposite leaves provide useful material for assays. A variety of schemes using such leaves have been described in which treatments (dilutions of inoculum), and leaf source are randomized statistically in such a way that as much undesirable variation can be excluded as possible.

An example of a suitable simple scheme using halved primary leaves of *Phaseolus vulgaris* might be as follows:

	Replicate							
Treatment	1	2	3	4	5	6	7	8
a	A1	B3	D2	E4	G1	H3	J2	K4
b	A2	B4	D3	F1	G2	H4	J3	L1
c	A3	C1	D4	F2	G3	I1	J4	L2
d	A4	C2	E1	F3	G4	I2	K1	L3
e	B1	C3	E2	F4	H1	I3	K2	L4
f	B2	C4	D1	E3	H2	I4	J1	K3

A–K are twelve separate plants from each of which four half leaves (1–4) are taken. Treatments 1–8 are sap dilutions. Using this and other similar local lesion hosts, detached leaves halved and kept moist on damp filter paper in enclosed trays may be inoculated.

Results of local lesion assays of a series of dilutions of one sap extract may be used to draw a dilution curve. Such dilution curves, which may be derived by manipulation of the data in a variety of ways [1], have a characteristic shape. At low concentrations of active virus the dilution curve approximates to a straight line which flattens at higher concentration. Comparisons of two separate inocula may be made with a reasonable degree of accuracy by assaying local lesion numbers within the lower range of dilutions which produce the straight line relationship.

Reliability of local lesion assays, factors affecting reliability and the statistical interpretation of results have been reviewed by Roberts [1].

Storage of Viruses

It is always valuable to retain type material of a range of viruses. Continuous culture of such viruses in host plants in the glasshouse is possible, but occupies valuable space and requires regular subculturing. Additionally, there is always the risk of contamination with other viruses and after repeated passages the virus may not retain its 'type' characteristics.

Alternative methods of storage of viruses are available but the success of these depends on the stability of the virus. Stable viruses may be stored in a variety of simple ways.

1 Entire systemically-infected leaves may be air dried and stored in suitable envelopes. This method is quite suitable for tobamoviruses.

2 Entire or partially shredded infected leaves may be stored in sealed polythene bags or stoppered bottles in the deep freeze at −20°C. Viruses remain infective for several months when stored in this way but preparations should not be thawed and refrozen. Store in small aliquots which can be removed and used up.

3 Expressed infective sap, centrifuged or partially clarified may be stored in the deep freeze at −20°C or for less stable viruses, in the refrigerator at 0–1°C. Some saps contain inhibitors which will reduce virus infectivity quickly. Clarified preparations do not keep as well as crude saps. As before, repeated freezing and thawing should be avoided.

Perhaps the most effective method of storing viruses, particularly those which are unstable, is by lyophilization or freeze drying. A variety of methods by which this can be achieved are available but depend to a large extent on the equipment available.

For relatively short-term storage and convenience of preparation, small cut sections of virus-infected leaf can be placed in ampoules, lyophilized and sealed in one simple process using the 'shelf and stoppering' accessory of the Edwards 'Modulyo' freeze dryer. Stoppered ampoules should then be stored in the refrigerator and not resealed after use. Decline in infectivity in such preparations varies but the procedure is relatively easy so that stocks may be regularly replenished.

An alternative but more time-consuming method of lyophilization has been described [2] in which some stored viruses have remained infective for more than 10 years. For this technique a centrifuge accessory for a freeze drier is required, but freeze drying may be by refrigeration alone or using phosphorus pentoxide as an additional dessicant.

Procedure

1 Select infected leaves when infectivity is maximal.

2 Grind in a pestle and mortar and squeeze the sap through cheesecloth.

3 To each 10 ml of sap add 7% w/v of d-glucose and 7% w/v peptone and mix.

4 Using a syringe with a wide bore needle place appropriate volumes of sap in freeze dry ampoules (0·25 ml in an 0·5 ml capacity ampoule). Take care not to wet the sides of the ampoule, since such streaks of sap on the tube side may become charred when heat-sealing ampoules and this may decrease the infectivity of the whole ampoule.

5 Primary drying under vacuum should be for about 6 hours, the ampoules being centrifuged for the first 5 minutes.

6 After primary drying, ampoules should be carefully constricted in a gas flame.

7 Constricted ampoules should be loaded onto the secondary drying apparatus, all unoccupied drying points blocked using empty ampoules.

8 Secondary drying after tube constriction takes a further 16 hours, after which tubes can be heat-sealed and labelled.

References

1. Roberts, D.A. (1964) Local-lesion assay of plant viruses. In *Plant Virology* (eds Corbett, M.K. & Sisler, H.D.), pp. 194–209. University of Florida Press, Gainesville.
2. Hollings, M. & Stone, O.M. (1970) The long-term survival of some viruses preserved by lyophilisation. *Annals of Applied Biology*, **65**, 411–418.

4

Transmission Tests

The recognition that symptoms caused by virus infections could be transmitted from one plant to another by transference of a '*contagium vivum fluidum*' was significant not only as the birth of the science of plant virology, but also in laying the foundations of what survives as one of the most useful of diagnostic tests. Early virologists found that not only could symptoms be transferred from one plant to another similar plant, but that other genera of plants could also be infected and, moreover, that the symptoms produced on these plants were characteristic for specific viruses. Thus, the sap transmission test with its use of a range of indicator plants (see Appendix, p. 88) was developed. Although as a diagnostic test the sap transmission test lacks the clinical objectivity and convenience of serological and electron microscopic techniques, it remains an essential component in the repertoire of the modern plant virologist. Sap transmission tests are still the best means of routine diagnosis for many viruses, the tests are sensitive and provide information essential in virus characterization. Sap transmission testing requires little equipment other than glasshouse accommodation and is therefore of considerable interest to the casual plant virologist. One obvious disadvantage of the test is its duration; reactions may take several weeks to occur. Additionally, for routine diagnosis, the need to keep many test plants available for use if necessary can be apparently wasteful of resources.

Not all plant viruses, of course, are sap transmissible and in the past it has often been necessary to resort to use of natural virus vectors to demonstrate virus presence by transmission test. The use of aphids as vectors in such transmission tests has become commonplace. The employment of other insects, nematodes or fungi is less commonly undertaken except in experimental circumstances.

Some virus diseases cannot be sap transmitted, nor can the natural vector be used in transmission tests, and for these the only means of confirmation may be by graft transmission test. In most cases the causal agents of such diseases are viroids or mycoplasma-like organisms, although sometimes merely the intransigence of the host plant demands a graft transmission test.

Unlike serodiagnosis, where specific reaction products indicate virus presence, or electron microscopy where virus particles are seen, virus diagnosis by transmission tests require a knowledge of a wide range of virus-induced plant reactions. The complete cataloguing of such reactions is impossible in such a limited volume (but see pp. 42–61) and the investigator must consult the appropriate literature [2]. In this chapter the techniques involved, the husbandry of test plants, vectors etc., and the precautions required in transmission testing are described.

Sap Transmission

Many viruses can be transmitted mechanically from suspect host plant to indicator or test plants (see Appendix, p. 89). Such transmission involves successful extraction of the virus from the host material and the transfer of the virus-bearing sap to the surface of the leaves of test plants in such a way that virus can enter cells. Provision must be made for indicator plant cells to be damaged, since in the absence of the natural vector, plant viruses cannot penetrate the cellulose cell wall. If the inoculated plant is susceptible, local lesions may appear on the inoculated leaves and systemic symptoms may develop on other parts of the plant. Experience of many years has shown that such reactions are characteristic for certain viruses or strains of virus and so can be used in their identification.

Materials

1 Host material suspected of being virus infected, usually leaves or petals showing symptoms, or occasionally roots or tubers. Young systemically invaded tissue usually contains the highest virus concentrations.

2* Test plants—selection of these is determined by the host plant under test.

3* Phosphate buffer 0·01 M pH 8·0.

4 Pestle and mortar (sterilized).

5 'Celite' or carborundum powder (600 mesh).

*See Appendix.

6 Labels and a pencil.
7 Cheese cloth, scissors.

Procedure

1 Thoroughly wash hands with soap and water (to avoid virus carry-over).

2 Select and label test plants, preferably at least two of each species, with date, host plant of origin of extract, reference number etc. On each plant select two leaves to be inoculated and mark each by pricking the tip with the pencil point (see Fig. 4.1). The two oldest, fully expanded healthy green leaves should be chosen for inoculation. Opposite leaves should be used if possible

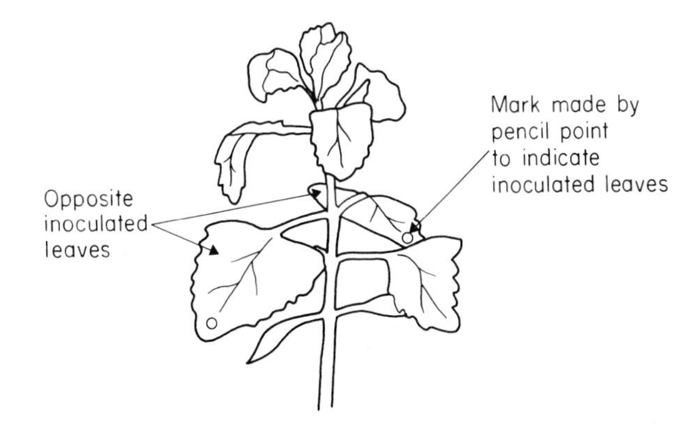

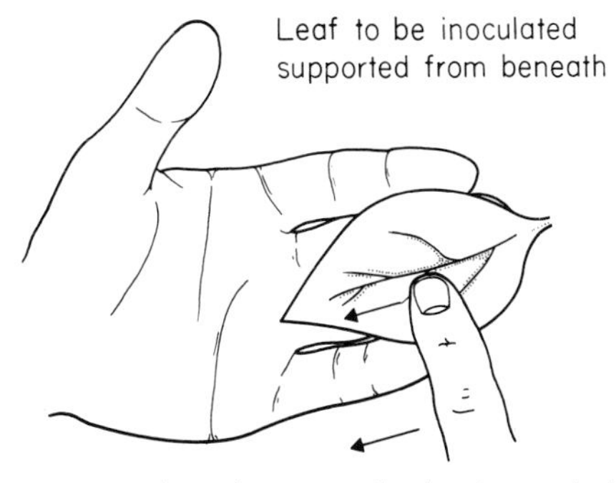

Fig. 4.1. Inoculation of *Chenopodium amaranticolor*.

and primary leaves should be avoided. If possible leaves of comparable age to that which is being tested may be chosen. When cucumber is used as a test plant the cotyledons are inoculated.

3 Select suitable infected material and place into the mortar. Large leaves may be cut into smaller pieces with sterile scissors if necessary.

4 Add 0·5–1 ml of phosphate buffer and an amount of 'Celite' equivalent to half the volume of a pea. (As an alternative to 'Celite' the indicator plant leaves may be dusted with carborundum powder as an abrasive. Care should be taken not to inhale carborundum dust).

5 Grind the leaf material until more or less homogeneous.

6 Squeeze the homogenate against the mortar edge to express liquid (or squeeze in cheese cloth).

7 Wet forefinger in the expressed sap and stroke gently onto each marked leaf, without delay, stroking from petiole to leaf tip whilst supporting the leaf with the other hand (see Fig. 4.1). Avoid heavy pressure and try not to go over the same area twice. (Where cross contamination of viruses presents a particular problem the pestle may be used to inoculate the leaf rather than the forefinger but this requires great care to avoid damaging the leaf surface.) A variety of other 'vehicles' have been used to inoculate the leaf: artists brushes, small sponges etc.; taking advantage of the sensitivity of the finger is, however, often best.

8 Immediately after inoculation wash inoculum off the test plant leaf, using a water jet from a squeeze bottle or suitably adjusted tap spray. Sap extracts left on the leaf reduce infection and may leave confusing residues.

9 Remove inoculated plants to the glasshouse section where they are to be grown on.

10 Collect together all equipment used for washing and sterilization later and finally wash hands again thoroughly before proceeding to the next test.

This basic procedure is suitable for many host virus combinations, although other methods of inoculation may be preferable in certain circumstances. For example, with viruses which are readily mechanically transmissible (e.g. TMV) the *detached leaf* test may be used. In this test, inoculum prepared as above, is rubbed onto detached leaves which are then incubated on damp filter paper within a Petri dish. This technique may be done in the laboratory reducing the risk of spread of virus by mechanical contact within the glasshouse. However, it is only applicable to viruses which cause local lesions within a few days of inoculation. For plants which contain inhibitors (see p. 31) the *dry inoculation* method gives good results. Leaves of test plants, having been marked as above are dusted with 'Celite' or carborundum powder. A suitable leaf from the infected plant is selected, folded several times then cut across the folds with a sharp knife. The cut surfaces are then rubbed gently over the leaves to be inoculated and finally test plants are washed. The dry inoculation method has been particularly successful with leaves of bulbous hosts.

Many mechanical devices have been described for the rapid inoculation of plants in routine diagnostic tests. However, the only one of these which offers any benefit other than speed or bulk handling is the artists airbrush. Using such equipment the virus inoculum, prepared as before can be sprayed on to test plants. The pressure of the spray can be adjusted so that sufficient leaf damage is caused by the abrasive (usually carborundum) in the preparation to allow infection. Use of an airbrush may give at least a tenfold increase in local lesion number with some viruses. On a large scale, spray gun application of viruses mixed in buffer containing carborundum has been used successfully for mass inoculation of crop plants in cross protection exercises.

Factors affecting susceptibility and infectivity

A variety of factors affect the success with which virus in sap extracts can be transferred to other plants and infect them. These factors may be concerned with the survival of the virus once removed from the plant, or with protection from toxic or inhibitory substances in the host sap. Preconditioning of indicator plants may increase their susceptibility and the conditions under which inoculated plants are kept may determine the extent of symptom development. Basic to all these predisposing factors is ability of the virus to infect the indicator plants chosen. Without some idea of the likely virus or viruses in a diseased plant the selection of indicator plants which are likely to be successfully infected may be fortuitous. However, a knowledge of virus symptomatology should provide some guidance and the literature [2] and sections in this volume (see pp. 42–61) may help. Having chosen a suitable indicator plant range the following should be taken into account.

Which part of the plant should be tested for virus?

Mechanically transmissible viruses are usually those which reach reasonably high concentration in their host and infect the epidermis as well as other parts of the plant. The mosaic mottle and ringspot viruses tend to be readily transmitted whilst those restricted in their distribution in the host, e.g. the luteoviruses, are not. Although choice of suspect material may often be limited, success is more likely if plants are chosen for test which have highest virus concentration. In inoculation studies it has been shown that for many viruses the concentration of particles reaches a maximum 2 or 3 days after inoculation. Virus concentration is often highest in young actively growing plant parts, which usually show the most pronounced symptom expression. Virus concentration in some plants may vary according to the season, e.g. Pelargonium leaf curl concentration is higher in the winter, and other environmental factors may have an effect, e.g. barley yellow mosaic concentration is highest during cooler times.

Choice of buffers for virus extraction

Phosphate buffers have been most widely used with success in sap transmission testing for many viruses. Such success is not due to the buffering action alone but to a phenomenon known as the 'phosphate effect'. Investigations have shown that phosphate increases infectivity, specifically on some test plants (e.g. *Phaseolus vulgaris* L.). However, not all viruses are affected in the same way, some being more successfully inoculated using alternative buffers. Lettuce mosaic virus is poorly transmissible using phosphate buffers but is transmitted effectively with borate buffer. In general, buffers with ionic strengths (0·05–0·1 M) and pH values (> 6·5) most suited to the cell are preferable. Such buffers preserve the virus during transmission and enhance the susceptibility of the test plant. The infectivity of most viruses is lost at low pH values.

Abrasives

The use of abrasives, either as a constituent in the suspect infected tissue maceration, or applied directly to the test plant leaf to be inoculated is vital for the sap transmission process. Whilst wounding of the cells of the test plant is essential this should not be so extensive as to mask virus symptoms. Thus, it is important to achieve a suitable balance, taking care in applying inoculum to avoid excessive vigour which is likely to be counter-productive. Simply rubbing the test plant with inoculum without added abrasive causes some wounding, leaf surface hair cells are broken and some virus infection takes place. However, experiments have shown that addition of 'Celite' may increase the number of successful infections by twenty to one hundred-fold. Use of abrasive is therefore essential; choice of which to use may be less important. Celite incorporated at tissue maceration tends to be less grossly damaging than carborundum applied to the test plant leaf, but much will depend on the experience of the operator and the toughness of the test plant leaf. Celite remains suspended in tissue macerates whilst carborundum quickly settles out. When abrasives have been used plants are liable to wilt and should therefore be kept in a humid, preferably shaded atmosphere for a few hours.

Inhibitors of infection

Host cell constituents released in tissue maceration may inhibit infection either by interference with the infection process or by inactivation of virus in the inoculum. Once infection is successfully achieved, these host cell constituents seem to have little action. Substances inhibiting infection occur in many plants and are of many different types. Proteins, polysaccharides and enzymes are most commonly implicated, but in many plants inhibitors have not been characterized. Ribonucleases, present in most if not all crude virus extracts, can significantly influence the infectivity of the virus and the susceptibility of the test plant. Whilst specific inoculum additives to overcome the effect of inhibitors can be used this is often not necessary in routine tests since, as inoculum is diluted, the effects of inhibitors diminish more rapidly than infectivity decreases. More frequently substances are added to the maceration buffer to prevent virus inactivation. Polyphenols in plant extracts are oxidized to quinones on release from the cell by polyphenoloxidase enzymes, and it is these oxidation products which inactivate virus. The action of polyphenoloxidases may be prevented by incorporation of reducing agents such as sodium sulphite, thioglycollic acid, mercapto-ethanol or ascorbic acid in the macerating buffer. Alternatively, chelating agents such as a diethyl dithio carbamate (DIECA) may be used.

Tannins are present in the sap from many woody plants (particularly Rosaceous hosts) and under certain conditions these may combine with, and precipitate, virus particles preventing infection. The effects of tannins may be overcome by preparation of extracts in buffers of high pH (8–9) or in buffer containing nicotine or caffeine, the combining action being greatly reduced in alkaline conditions. Alternatively protein such as hide powder or adsorbents such as polyvinyl pyrollidone may compete with the virus particles for the tannin in the extraction medium.

Substances which prevent successful sap transmission of viruses occur in greater concentration in some plant parts than in others. Young leaf material usually has less inhibitor activity than older leaves and petals have least of all. The action of inhibitors may be overcome under certain circumstances by using the dry inoculation method described above.

Test plant physiology

The susceptibility of a test plant to infection depends greatly on its physiological state. Plants grown in shade conditions with pale green juicy leaves are more susceptible than those grown in high light conditions with tough dark green leaves containing less water. Nutrient conditions that encourage rapid plant growth should be adopted and growing temperatures should be comparatively high. Changes in test plant environment immediately before inoculation often significantly increase the chance of success. The most consistently effective action is to darken plants for a period of up to 2 days before inoculation. The effect of darkening is complex and may be modified by day-length prior to the darkening period, or exposure to light after darkening but before inoculation. Keeping test plants in warm air before inoculation increases their susceptibility, as does wilting of test plant leaves.

After inoculation the prompt washing off of excess inoculum is beneficial and test plants should then be grown-on in cooler light conditions.

Recording the result

Observation of the symptoms which develop on inoculated test plants is of paramount importance in the sap transmission test if comparative results are to be obtained. Whilst interpretation of symptoms may be considered to some extent subjective, certain conventions help to encourage comparative records. Observe the inoculated test plants regu-

Fig. 4.2. Local pin-point chlorotic lesions caused by lettuce mosaic virus on *Chenopodium amaranticolor*.

larly (preferably daily) and make descriptions of symptoms and note when they are first seen. Distinguish between 'local' reactions on the inoculated leaves (see Figs 4.2 and 4.3), which may occur only 2 days after inoculation, and systemic symptoms on the non-inoculated leaves (see Fig. 4.4), which generally take longer to show.

(a) Local lesions: record the diameter, colour at the centre and at the margin, define as chlorotic (pale green) or necrotic (dead). Note any progressive enlargement of local lesions on successive days. Local reactions may also sometimes cause ringspots.

(b) Systemic symptoms: systemic reactions may be much more variable than local reactions. Necrotic spots or ringspots may occur but more often systemic reactions show as mosaics or mottles or symptoms associated with the veins. Vein clearing or vein banding may be chlorotic or necrotic, and give line, ring, or oak leaf patterns. Distortion, tendril formation (see Fig. 4.5), bubbling of leaf surfaces or the production of outgrowths or enations may occur. The growing point of some test plants may be severely distorted or killed by some viruses.

Fig. 4.3. Necrotic ringspots: 'local' lesions caused by tobacco rattle virus on *Nicotiana tabacum*.

Problems

A number of factors may complicate the interpretation of sap transmission test results. Physical damage to the test plant inflicted during inoculation can be confused with some local virus reactions. Generally this is a problem overcome with experience but inoculation of test plants with healthy sap as controls should help in difficult situations.

Latent viruses present in test plants may become apparent during tests. Specifically, sowbane mosaic virus may be seed-borne in the

Fig. 4.4. Systemic mosaic caused by invasion of new leaf *Cucumis sativus* by cucumber mosaic virus.

otherwise healthy test plant *Chenopodium quinoa* but may not produce symptoms. Abrasion of such plants during sap transmission tests may stimulate the development of local and systemic reactions of this virus. The problem can be overcome by inoculating a few test plants to each other. In the absence of symptoms these should be kept for seed, but other *C. quinoa* plants should not be allowed to flower to avoid pollen transmission of the virus. Accidental mechanical transmission of sowbane mosaic may also occur and should be avoided with care.

Extreme growing temperatures after inoculation may mask symptoms or prevent development of systemic reactions; e.g. TMV produces only local lesions on the test plant *Nicotiana glutinosa* under cool conditions, but may become systemic as temperatures rise.

Mixtures of several viruses may cause confusing symptoms in inoculated test plants. Mixtures may often be separated by inoculation to a wider range of test plants, some of which may not be susceptible to one component virus. Additionally, aphid or other vector transmission (see below) may isolate one virus.

Contaminant viruses, particularly those which are very readily mechanically transmitted, may complicate test plant reactions. Careful

Fig. 4.5. Severe systemic distortion of *Nicotiana glutinosa* leaves by tomato aspermy virus.

and systematic washing at relevant stages in the inoculation process is essential to avoid cross contamination. Strict hygiene practices in the virus glasshouse should be observed to avoid the introduction of contaminant viruses on clothing etc., and smoking is usually forbidden to avoid introduction of tobacco mosaic virus. Testing for particularly contagious viruses should be in separate glasshouses or as detached leaf tests in isolation (see p. 28).

Vagrant aphids or other insects may cause confusing damage or spread viruses within the glasshouse and should be kept out at all costs (Table 4.1).

Raising test plants

Most test plant seed can be sown in seed trays or pots in a proprietary seedling compost, e.g. Levington's Universal. Seeds should be sown evenly (it is easier to sow small seeds evenly if mixed with a little silver sand) and covered with a thin (2 mm approximately) layer of fine compost. Water thoroughly by gentle spraying from above. Large seeds, e.g. cucumber, bean and pea, should be sown in sand.

After sowing, seeds can be left to germinate in a forcing box or propagator. Suitable small propagator units with heating wire, thermostat and transparent cover are available from horticultural suppliers. The soil temperature in the propagator should be kept at 27°C with air temperature about 22–24°C. Seed boxes, before and after germination, should not be watered more than once a day while in the propagator.

Plants with large seeds, and *Chenopodium amaranticolor* should be removed from the propagator on germination (seedlings of the latter are killed by heat). On removal from the propagator, prior to pricking-out, seedlings should be kept well watered in a temperature of 18–21°C. Humidity in the glasshouse should be kept high, especially in summer, by frequent damping down of the floors. Seedlings should be pricked out into pots (clay pots are best but plastic pots are less expensive and are satisfactory) as soon as they are large enough to handle, paying careful attention to hygiene to avoid accidental inoculation with virus. Seedlings become established more quickly if kept in humid conditions.

Plants with special requirements

Chenopodium amaranticolor. Seedlings emerge in flushes so it is advisable to sow more than required, use the first flush and discard the remainder. Remove pots of seedlings from the propagator on germination, high temperature kills young seedlings. Prick out into pots at the appearance of first true leaves, and do not water too heavily or roots will not develop well. Grow on at 13–18°C. Supplementary lighting to maintain summer day-length prevents flowering (plants with flowers are unsuitable for testing).

Datura stramonium. Encourage germination by scratching or removing testa, or by freezing for 7–10 days. The seed coat tends to stick to the seedling and should be removed to avoid distorted growth.

Gomphrena globosa. Seedlings should be pricked out into pots at the advanced cotyledon stage. This plant grows slowly in winter and may need up to 14 weeks from sowing before suitable for inoculation. Grow at 18–21°C.

Nicotiana clevelandii. Dormancy in seeds must be broken by soaking seed for 2 hours in giberellic acid solution (1 g litre^{-1}) before sowing. Water sparingly to avoid damping off.

Nictoniana glutinosa. Prone to damping-off. Seed should be sown thinly and watered sparingly. Prick out promptly and give good light conditions after pricking out to aid establishment. Pricking-out into peat blocks as an intermediate stage before final potting is said to promote shorter growth. Grow at 13–18°C.

Tetragonia expansa. Pot seedlings on appearance of first true leaves. Growth is slow in winter and plants become straggly if not potted deep enough.

Plants with large seeds, e.g. cucumber, pea, bean, should be potted up when cotyledons separate.

Glasshouse management

Glasshouse management should be such as to produce vigorously growing soft lush plants. Particular attention should therefore be paid to ventilation, shading, watering, lighting and hygiene.

Ventilation. Glasshouses for virus work should be aphid proofed and this may create problems of ventilation. Ridge and wall ventilators may be supplemented by thermostatically controlled fans which draw air into the house from the outside. Such forced ventilation provides additional cooling capacity when outside ambient temperatures are not too high and may be arranged so as to promote cooling by evaporation of water from glasshouse floors. Fan inlets should of course be aphid proofed. Modern air conditioning units are an ideal solution to this problem.

Shading. Shading roller blinds, together with 'summer cloud' on south-facing walls should be used as necessary to minimize summer extremes of temperature.

Heating. Thermostatic controls should be set to maintain a temperature of 18–21°C.

Watering. Regular watering should be arranged so as to avoid water stress in test plants, and to avoid over watering. Capillary sand benches are useful if properly maintained, and capillary mat systems may be a satisfactory alternative; however, some viruses spread in drainage water (tombusviruses) and others have fungal vectors which do so. If these are to be encountered, appropriate precautions should be taken. Staging and floors should be periodically damped down during summer days to keep a humid atmosphere.

Lighting. Supplementary lighting is essential for proper plant growth during the winter months: 16–18 hours artificial illumination should be given to overlap daylight. Individual circumstances may dictate the kind of lighting equipment. However, for general use the MBF/U mercury vapour lamp has been found suitable.

Recent experience in potato tuber testing has suggested that the light output of high pressure sodium lamps is of a suitable quality for good plant growth and enables these to be used for illuminating large areas. Such lamps may be housed in fittings which enclose all the electrical components (except time switches) allowing them to be plugged into existing sockets. Alternatively, less bulky fittings require separate ancillary equipment.

The choice of reflectors depends very much on the type of lamp used, their siting, and distance from the plants.

Seed saving and supply

Many, but not all, seeds for test plant raising are available from commercial suppliers. However, the test plant reactions described in the literature usually refer to specific cultivars or breeding lines. Thus, saving seed of such cvs and lines may be more convenient and indeed may be the only way to maintain stocks of some genera. Test plants for which specific cvs are recommended are:

Chinese cabbage	cv. 'Pe tsai';
Cucumber	cv. 'Butcher's Disease Resister';
Vicia bean	cv. 'Seville Long pod';
Phaseolus bean	cv. 'The Prince';
Pea	cv. 'Onward'.

Specific lines of *Nicotiana* species have found frequent use amongst virologists and may only be available from virology laboratories. The reactions of such 'lines' are different with certain viruses and thus the lines are valuable as distinct test plants. Seed of such *Nicotiana* spp., e.g. *N. tabacum* 'White Burley' and *N. tabacum* 'Samsun NN' and other *Nicotiana* species must be saved.

Other seed types of unspecified cv. are available from commercial seedsmen, e.g. *Gomphrena globosa* (a mixed selection is satisfactory), and *Petunia hybrida*. For those intending to undertake regular testing it is important to save seed of the other regularly used test plant species. 'Clones' of plants showing desirable test plant characteristics can be specifically selected and retained in this way, e.g. broad entire-leaved *Chenopodium* spp.

Seed collection is usually best carried out in the autumn. Several plants to be kept for this purpose should be set aside, exposed to normal autumn day-length and allowed to flower and set seed naturally. For most of the plants it is simply necessary to wait until flowering is over and the seed heads are dry. Seed can then be extracted by rubbing the heads inside a paper bag. *Physalis* produces its seed 'berry' inside a green 'lantern'. This must be removed as seeds mature and berries turn brown. *Datura* 'thornapples' should be removed when fully grown, left to dry then split to obtain seed.

Pest and disease control

Quarantine

Whenever possible a quarantine period should be given to plant material before it is taken into the glasshouse. Any showing symptoms of pests or diseases should be rejected.

Glasshouse hygiene

Plant debris should be removed from the glasshouse as soon as possible. Weeds should not be allowed to develop either inside or immediately round the glasshouse. If left uncleared, weeds or plant debris act as reservoirs of pests and diseases or may provide an intermediate virus host.

Pests

Since many insect pests are potential virus vectors it is essential to keep them out of the virus glasshouse. Insects may enter the glasshouse through doors and ventilators and on clothing. Insect proofing can rarely be totally effective and a programme of insecticides is an essential supplement (Table 4.1). Care should be taken to avoid keeping any plants beyond their useful life since it is these on which pest populations are likely to build up.

Resistance
Some strains of the major glasshouse pests have become resistant

to chemicals that have been used in the past to control them. To reduce the chance of development of further resistant strains, pesticides from different chemical groups should be alternated in pest control programmes (Table 4.1).

Methods of treatment

Common methods of insecticidal treatment are the use of smokes, fogs, low volume sprays (aerosols) and high volume sprays. The

Table 4.1. Chemicals available for pest control

		Pest			
Pesticide group	Active ingredient	Aphid	White-fly	Two-spotted spider mite	Sciarids
Carbamate	pirimicarb	✓(S,WP)	X	X	X
Pyrethroid	deltamethrin	✓(EC)	✓(EC)	X	X
Organochlorine	HCH*	✓(EC,S)	✓(S)	X	X
	dienochlor†	X	X	✓(WP)	X
	dicofol/tetradifon mixture‡	X	X	R(EC)	X
Organophosphorus	demeton-s-methyl	R(EC)	X	R(EC)	X
	heptenophos	✓(EC)	X	X	X
	pirimiphos-methyl	X	✓(F)	X	X
	dichlorvos§	R(EC)	✓(EC)	R(EC)	✓
	cyhexatin‖	X	X	✓(WP)	X
	petroleum oil**	X	X	✓(L)	X
	nicotine	✓(Solution)	X	X	X
	diflubenzuron	X	X	X	✓

✓, May be used and should be effective.
R, May be used, but some insect strains show resistance.
X, Not effective.
(S), Smoke.
(EC), Emulsifiable concentrate.
(WP), Wettable powder.
(L), Liquid.
(F), Fog.
*Do not use on cucumber. Will not control mottled arum aphid.
†Do not use on edible crops.
‡Do not use on seedlings, young plants or cucumber before mid-May.
§Do not use on cucumber or at temperatures below 15°C.
‖Do not use on cucumber or within 28 days of spraying oil.
**Do not use within 28 days of sulphur or cyhexatin sprays.
If using a spray for the first time *try on a few plants first.*
N.B. The above table has been prepared as a guide only and does not claim to be exhaustive.

latter are more efficient since good cover is guaranteed if properly done. However, the other methods of treatment may be more simple, and do not leave unsightly deposits on plant tissues.

Side effects

Certain chemicals may produce undesirable phytotoxic effects on some subjects. Flowering parts are particularly susceptible. For this reason, if a chemical is to be used on an unusual subject it should be tried on a few plants first and time allowed for any symptoms to develop before the rest of the plants are treated.

Diseases

Diseases are, by comparison, less troublesome than pests in the virus glasshouse. Few leaf pathogens present problems and usually, apart from normal hygiene precautions, routine prophylactic control programmes are unnecessary. The diversity of plant genera in the virus house may discourage the build up of leaf pathogens, and regular disposal of plants when their useful life is over will help. Tobacco blue mould (*Peronospora tabacina*) has occasionally caused problems in *Nicotiana* species in virus glasshouses in the past.

Symptoms

(a) The first sign of infection is a slight yellowing of affected leaves, followed by a diffuse yellow coloured area halfway up the shoulder of the plant 2–3 days later.
(b) Under favourable conditions, the fungus sporulates on the under surfaces of the yellowed areas after about a week.
(c) Later, the chlorotic areas on affected leaves become necrotic. Necrosis is more common on infected tobacco grown out of doors rather than on plants grown under glass.
(d) The fungus sometimes becomes systemic in the plant, causing a dwarfing and yellowing in *N. glutinosa*.
(e) On infected seedlings, the fungus sporulates on both surfaces of the leaves.

Control

Whilst modern fungicides (dithiocarbamate or metalaxyl) may effectively limit the spread of this disease it is advisable to take additional stringent eradication measures if it occurs.

All infected and clean tobacco plants in the glasshouse should be destroyed by immersion in a bath of formalin. All clay pots and soil used in them should be thoroughly steam sterilized for at least 2 hours. Glasshouse cubicles should be 'cleaned-out' and the whole structure drenched with 2% formalin. Duck boards, if present, should receive an extra strong formalin drench. Each cubicle should receive this treatment three or four times at about 5-day intervals. Tobacco should *not* be grown for 3 weeks at least.

During the period after an outbreak, sprays of fungicide (dithiocarbamate or metalaxyl) may help to prevent further attacks. However, routine sprays are not normally necessary.

Powdery mildew (*Erysiphe graminis*) often attacks oat test plants as they mature in the glasshouse. This can be prevented by seed treatment with ethirimol or triadimenol. Alternatively, sprays (one or two) of triadimefon have been effective. Sprays of fungicide can cause scorching of the softer glasshouse-raised plants and, since some fungicides have undisclosed aphicidal properties when used as sprays, it is wise to avoid treatment before transmission feeding is completed when plants are used for vector transmission testing (see below).

Root-infecting fungi are normally of little concern as pathogens. *Pythium* species may build up in capillary mat systems if these are neglected, but normally present no problem when plant turnover is sufficiently rapid. Root-infecting vector fungi (*Olpidium, Polymyxa,* etc.) are obviously of greater concern and care should be taken to avoid their introduction. Use of proprietory potting composts or sterilized loam should avoid this problem.

Preliminary routes for virus identification by sap transmission

Whilst texts describing diagnostic methodology are not uncommon, keys providing routes for virus identification are almost nonexistent. The great variety of reactions given by different viruses under different conditions and with different indicator plants perhaps explains this lack of information. The following table is an attempt to direct the inexperienced diagnostician along lines of investigation using sap transmission tests only to give identification of a limited range of viruses. The key requires as its first step, the inoculation and observation of symptoms on *Nicotiana tabacum* 'White Burley'. Further distinction of different viruses is achieved by a supplementary inoculation to other named test plants where appropriate. The reactions described are those which normally occur under suitable glasshouse conditions and where the initial inoculum contains a sufficient virus concentration, but strain variations make logical sequence progression in key form difficult. Viruses are listed according to the type of reaction they produce in *N. tabacum.*

Local symptoms in tobacco	Systemic symptoms in tobacco	Virus	Confirmatory host(s)	Symptoms on confirmatory host
(a) Necrotic local lesions and no systemic reaction				
Necrotic lesions 2 mm diam. (occasionally systemic mosaic, see below) (variable according to strain.		Tomato mosaic (some strains) (TMV)	*N. glutinosa*	Grey or brown spots 2–3 mm diam. enlarging with a dark brown edge.
Coarse circular brown blotches or rings up to 10 mm diam., few in number.			*P. vulgaris*	Pin-point necrotic local lesions in 1–3 days; not systemic.
Coarse extensive etched pattern. Scattered necrotic spots gradually forming a pattern in relation to main vein (variable according to strain).		Tobacco rattle (TRV)	*C. amaranticolor*	Necrotic local lesions, some tending to spread; not systemic.
Greyish brown spots 1–5 mm diam. often surrounded by broken rings up to 10 mm diam. Sometimes dart-shaped necrotic banding of main veins.		Tobacco necrosis (TMV)	*P. vulgaris*	Discrete red local lesions, or local lesions spread along minor veins up to 3 mm. One strain gives systemic invasion.
Round spots up to 2 mm diam. first reddish then pale sometimes with a brown edge.		Tomato bushy stunt (type strain) (TBSV)	Tomato	Chlorosis and necrosis of leaves.
			Datura	Local lesions then bright yellow and green mottle.
			C. amaranticolor	Round grey spots up to 1 mm diam. often with a narrow chlorotic edge.

Spots, rings or blotches enlarging, possibly covering much of leaf. Local necrosis may extend to dart-shaped necrotic banding of virus.		Tomato spotted wilt (some strains) (TSWV)	*N. glutinosa*	Local lesions as with *N. tabacum* but followed by systemic necrotic patterns.
Chlorotic spots, up to 5 mm becoming centrally necrotic, brown rimmed with a chlorotic halo.		Turnip mosaic (TuMV)	*C. quinoa*	Chlorotic and necrotic local lesions and systemic veinal flecks and spots

(b) Necrotic lesions followed by faint systemic chlorosis

The common nepoviruses may give faint or conspicuous local symptoms followed by systemic invasion, which may be evident only as vague chlorosis or may be more definite; see (d).

(c) Necrotic lesions followed by systemic necrosis only

Spots, rings or blotches becoming larger necrotic dart-shaped vein banding; see (a).	Zonate lesions, necrosis of stem and growing point.	Tomato spotted wilt (some strains) (TSWV)	*N. glutinosa*	Local lesions followed by systemic necrotic patterns.

(d) Necrotic or necrotic and chlorotic local lesions followed by systemic chlorosis with necrosis

Coarse spots up to 2 mm diam., grey/brown with a brown edge.	Scattered zonate chlorotic blotches or necrotic spots with chlorotic halo.	Arabis mosaic (AMV)	*C. quinoa*	Spots and fine rings up to 1 mm diam., chlorotic or necrotic with a halo. Faint chlorotic systemic streaking or line marks and distortion.
Grey to white ringspots up to 2 mm diameter.	Faint chlorotic rings and spots.	Raspberry ringspot (RRV)	*C. amaranticolor*	Chlorotic or necrotic local lesions; no systemic symptom.
Chlorotic or necrotic spots or rings.	Spots, rings and line patterns with variable necrosis.	Tomato black ring (TBRV)	*C. amaranticolor*	Chlorotic or necrotic local lesions, systemic necrosis or chlorotic mottle.

Local symptoms in tobacco	Systemic symptoms in tobacco	Virus	Confirmatory host(s)	Symptoms on confirmatory host
Grey-white rings up to 3 mm diam. or 5 mm on young leaves with wide chlorotic halo.	Systemic leaves first infected may be distorted with grey lesions with brown margins up to 5 mm diam., sometimes generalized necrosis, ring or line patterns.	Tobacco ringspot (TRSV)	*C. quinoa*	Local necrotic dots, not systemic.
			Cucumber	Local chlorotic lesions and systemic mottling and dwarfing and apical distortion.
Chlorotic or finely etched necrotic ring spotting	Fine vein banding and ring spotting, tortoise-shell pattern with necrotic etching; later leaves 'grow out' of symptom particularly at high tempera-tures.	Potato X (some strains see (f)) (PVX)	Datura	Chlorotic rings turning to a mosaic mottle.
			Gomphrena globosa	Local necrotic lesions with a red margin.
Coarse watermarking over lamina with necrotic arcs and lines.	General chlorosis with coarse arcs and ringspots up to 4 mm, mild partial veinal necrosis and leaf distortion.	Tobacco rattle (see (a)) (TRV)	*C. amaranticolor*	Necrotic local lesions spreading but not systemic.

Local chlorotic or necrotic lesions up to 2 mm diam. followed by systemic vein clearing of the potato virus Y type (see (f)) may be due to tobacco etch

(e) Chlorotic local lesions followed by obvious systemic necrosis and some chlorosis

Faint chlorotic or necrotic spots up to 3 mm, general chlorosis as virus becomes systemic.	Fine vein clearing rapidly becoming a fine vein necrosis spreading to main veins and midrib. Affected leaves curl and crumple.	Potato virus Y (tobacco veinal necrosis or Y^N strain) (PVY^N)	Chenopodium	Eventual reddish local lesions.
			Datura	No reaction.

The nepoviruses when in low concentration may produce few local lesions but systemic rections should be evident.

(f) Chlorotic local lesions followed by systemic chlorosis

Vague spots up to 3 mm diam. sometimes necrotic, arcing or rings.	Diffuse green vein banding of finer veins mostly towards leaf tips. Possibly accompanied by chlorotic dots and rings.	Potato virus X (PVX)	Datura	See (d) for reactions.
			Gomphrena globosa	
Faint chlorotic spots up to 3 mm.	Vein clearing followed by more or less obvious mottling which may later disappear. Veinal necrosis indicates Y^N strain; see (e).	Potato virus Y (PVY)	*Chenopodium*	Eventual reddish local lesions.
			Datura	No reaction.

Local symptoms in tobacco	Systemic symptoms in tobacco	Virus	Confirmatory host(s)	Symptoms on confirmatory host
Possible vague round chlorotic spots up to 3 mm.	First non-local leaf symptomless or vaguely chlorotic, later leaves have distortion and dark green vein banding or blistering.	Tobacco mosaic (some strains) (TMV)	*N. glutinosa*	Grey or brown spots 2–3 mm diam. enlarging with a dark brown edge.
Vague ill-defined chlorotic spots up to 2 mm sometimes necrotic spotting, arc or ring banding chlorotic spot.	On first non-local leaf chlorotic spots up to 5 mm diam. Later leaves chlorotic mottle/vein banding. Plants eventually become symptomless.	Cucumber mosaic (CMV)	Cucumber	Local lesions followed by systemic mosaic.
			Chenopodium	Numerous sharp dots, pale becoming necrotic.
			N. glutinosa	Local chlorotic spots with systemic mosaic and some distortion
As above (CMV).	At first as above (CMV). First systemically-invaded leaf may show zonate water marking. Later leaves dark green blister like banding sometimes giving an oak-leaf effect.	Tomato aspermy (TAV)	Cucumber	Possibly some local lesions but no systemic invasion.
			Chenopodium	As for CMV.
			N. glutinosa	Severe systemic distortion with leaves eventually reduced to tendril forms.

Indicator plants used: Cucumber, *Cucumis sativus*; *Datura*, *Datura stramonium*; *Chenopodium*, *Chenopodium amaranticolor* or *C. quinoa* or as

Viruses on crop plants

Test plant reactions for specific viruses on a small range of common host crop plants.

	Nicotiana tabacum (cv. White Burley)		*Nicotiana glutinosa*		*Chenopodium amaranticolor*		Other test plants
	Local	Systemic	Local	Systemic	Local	Systemic	
Lettuces							
Dandelion yellow mosaic virus	No infection		No infection		Rather faint, local chlorotic spot lesions after 1–15 days, enlarging up to 2–2·5 mm diameter after about 3·5 weeks, and sometimes becoming necrotic		
Alfalfa mosaic virus	Local chlorotic spots after 3–6 days	Systemic vein clearing, chlorotic spots or rings developing to a ring mottle. After about 3 weeks subsequent growth shows few symptoms or none	Local chlorotic or grey spots after about 3 days	Systemic vein clearing, mottle and distortion follows in 1–2 weeks. Symptoms continue to develop on subsequent growth for some months	Faint semi-necrotic local dots after 4–6 days	Rapid systemic invasion with chlorotic flecks streaks and spots severe leaf curling and buckling, and a grey, mealy appearance of the young leaves	*Phaseolus vulgaris* (cv. The Prince). Local: numerous local discrete red-brown dots after 2–3 days; these do not extend along adjacent veinlets. When lesions are very numerous inoculated leaves soon collapse and fall off. Systemic: systemic mottle frequently follows with variable leaf distortion and rugosity though later young growth may be symptomless.

	Nicotiana tabacum (cv. White Burley)		*Nicotina glutinosa*		*Chenopodium amaranticolor*		Other test plants
	Local	Systemic	Local	Systemic	Local	Systemic	
Tomato spotted wilt virus	Local lesions variable; usually necrotic spots, rings or irregular local lesions after 3–7 days, usually enlarging considerably and sometimes leading to total necrosis of inoculated leaves	Many isolates become systemic, causing necrotic flecks, rings or veinal necrosis which may kill the plant	Large, local necrotic spots after 4–9 days; sometimes local lesions are initially small, but soon enlarge	Most isolates become systemic, with necrotic flecks and streaks, often along the small veins. Necrotic areas may develop; apical necrosis may kill the plant	Local necrotic dots or ring-spots after 3–7 days		
Cucumber mosaic virus	Local chlorotic (or necrotic) spots after 3–7 days	Systemic vein clearing, mottle and distortion of very variable colour and intensity with different strains; severe strains lethal in winter, mild strains may show very slight mottle on one young leaf followed by symptomless, young growth	Similar variable mottle and leaf distortion. Usually, a given strain causes more severe symptoms on this species than on tobacco		Numerous local pale green or whitish dots after 3–6 days		*Cucumis sativus* (cv. Butcher's Disease Resister). Different strains cause variable mottle, leaf distortion and stunting

Lettuce mosaic virus	No infection		No infection		bright green spots after 6–12 days, sometimes absent	spots, veinal flecks or yellow-netting of the young leaves. Though readily isolated from youngish plants, it is difficult to transmit the virus from mature lettuces	Whitish, local necrotic dots after 3–8 days enlarging during the next 1–2 weeks to form maroon-edged rings. In the early stages these lesions are often better seen from the underside of the leaf
Arabis mosaic virus	Local necrotic or chlorotic spots	Systemic invasion often occurs, with yellow rings and line patterns and variable distortion	Local grey sunken spots after 7–9 days; sometimes no local symptoms	Variable systemic chlorosis and ring-mottle	Local chlorotic or semi-necrotic small spots after 5–7 days, sometimes absent	Systemic yellow-brown flecks after 8–14 days with variable necrosis, distortion and wilt	*Cucumis sativus* (cv. Butcher's Disease Resister). Incubation period may be more than three weeks. No local lesions. A mosaic mottle *Phaseolus vulgaris* (cv. The Prince). Faint local chlorotic irregular spots; systemic variable mottle, with distortion and necrosis
Leguminous plants							
Cucumber mosaic virus	Local chlorotic (or necrotic) spots after 3–7 days	Systemic vein clearing, mottle and distortion of very variable colour and intensity with different strains; severe strains lethal in winter, mild strains may show very slight mottle on one young leaf, followed by symptomless young growth	Similar variable mottle and leaf distortion. Usually, a given strain causes more severe symptoms on this species than on tobacco		Numerous local pale green or whitish dots after 3–6 days		*Phaseolus vulgaris* (cv. The Prince). Tiny local necrotic dark specks after 7–10 days, in the winter months only: these lesions sometimes need a hand lens to be identified with certainty *Cucumis sativus* (cv. Butcher's Disease Resister). Different strains cause variable mottle leaf distortion and stunting

	Nicotiana tabacum (cv. White Burley)		*Nicotina glutinosa*		*Chenopodium amaranticolor*		Other test plants
	Local	Systemic	Local	Systemic	Local	Systemic	
Tomato spotted wilt virus	Local lesions variable; usually necrotic spots, rings or irregular local lesions after 3–7 days, usually enlarging considerably and sometimes leading to total necrosis of inoculated leaves	Many isolates become systemic, causing necrotic flecks, rings or veinal necrosis which may kill the plant	Large, local necrotic spots after 4–9 days; sometimes local lesions are initially small, but soon enlarge	Most isolates become systemic, with necrotic flecks and streaks, often along the small veins. Necrotic areas may develop; apical necrosis may kill the plant	Local necrotic dots or ring spots after 3–7 days		
Alfalfa mosaic virus	Local chlorotic spots after 3–6 days	Systemic vein clearing, chlorotic spots or rings, developing to a ring mottle. After about 3 weeks, subsequent growth shows few symptoms or none	Local chlorotic or grey spots after about 3 days	Systemic vein clearing, mottle, and distortion follow in 1–2 weeks. Symptoms continue to develop on subsequent growth for some months	Faint, semi-necrotic local dots after 4–6 days	Rapid systemic invasion, with chlorotic flecks, streaks and spots, severe leaf curling and buckling, and a grey, mealy appearance of the young leaves	*Phaseolus vulgaris* (cv. The Prince). Numerous local discrete red-brown dots after 2–3 days; these do not extend along adjacent veinlets. When lesions are very numerous inoculated leaves soon collapse and fall off. Systemic mottle frequently follows, with variable leaf distortion and rugosity, though later young growth may be symptomless *Vicia faba* Local reddish-brown dots after 3–6 days. Many plants rapidly develop systemic stem necrosis and a lethal wilt

	Nicotiana tabacum (cv. White Burley) and *Nicotiana glutinosa*	*Chenopodium amaranticolor*		*Phaseolus vulgaris*		*Vicia faba*	
		Local	Systemic	Local	Systemic	Local	Systemic
Leguminous plants (cont.)							
Bean yellow mosaic virus	No infection	Few faint chlorotic local spots after 2 weeks, becoming necrotic; not reliable		Faint, local chlorotic dots and spots after 7–12 days	Severe spot mottle after 2 weeks Symptoms may later become more severe and include severe leaf down-curling, stunting and mottle	Prominent local reddish-brown irregular small spots and rings after 7–10 days.	Systemic chlorotic flecks, spots and mild mottle after 2–3 weeks
Broad bean mottle virus	No infection	Numerous local chlorotic dots after 5 or 6 days; this reaction occurs only under favourable conditions			Variable systemic mottle		Systemic vein-clearing followed by a bright inter-veinal mottle. Extensive necrosis in winter
Pea enation mosaic virus	No infection	Not tested		No infection			Systemic spot mottle and irregular veinal flecks after 2–4 weeks; characteristic glassy flecks—the so-called 'windows'

	Nicotiana tabacum (cv. White Burley) and *Nicotiana glutinosa*	*Chenopodium amaranticolor*		*Phaseolus vulgaris*		*Vicia faba*	
		Local	Systemic	Local	Systemic	Local	Systemic
Pea mosaic virus	No infection	Local chlorotic spots after 1–2 weeks	Systemic chlorotic spots, veinal flecks and often bright yellow-netting of the veins, after a further 1–3 weeks	No infection			Systemic veinal mottle after 2–4 weeks, often developing to a characteristic 'chevron' pattern across the midrib

	Nicotiana tabacum (cv. White Burley)		*Nicotiana glutinosa*		*Chenopodium amaranticolor*		
	Local	Systemic	Local	Systemic	Local	Systemic	Other test plants
Cruciferous plants							
Turnip mosaic virus	Local chlorotic spots after 4–7 days, each later developing a chlorotic halo and central necrotic area		Local chlorotic and semi-necrotic spots after 5–9 days	Fully systemic rugose mottle, vein clearing, often with distortion and patches of leaf necrosis. Severe	Local chlorotic spots after 5–11 days, becoming red-rimmed rings 2–3 weeks, and characteristically extending slightly along adjacent veins		
Cauliflower mosaic virus	No infection		No infection		No infection		*Brassica pekinensis*. Systemic vein clearing followed by a coarse mottling
Turnip crinkle virus	No infection		No infection		Local chlorotic dots after 5–7 days		*Brassica pekinensis*. Systemic leaf crinkling and some dwarfing
Turnip rosette virus	No infection		No infection		No infection		*Brassica pekinensis*. Yellow local lesions followed by slight systemic crinkle andvein necrosis
Turnip yellow mosaic virus	No infection		No infection		No infection		*Brassica pekinensis*. A bright yellow veinal mottle

	Nicotiana tabacum (cv. White Burley)		*Nicotina glutinosa*		*Chenopodium amaranticolor*		Other test plants
	Local	Systemic	Local	Systemic	Local	Systemic	
Potatoes							
Potato virus X	Variable local chlorotic, necrotic or semi-necrotic spots and rings after 5–8 days; symptoms vary greatly	Mild strains may cause only slight systemic spot-mottle followed by vein banding of the young leaves. Severe strain may induce considerable necrosis, severe mottle, necrotic veinal flecks, distortion and stunting	Local chlorotic spots	Variable systemic speckle-mottle, or more severe distorting mottle and necrosis	Numerous pale green, or fawn, local necrotic dots after 5–7 days; lesions from some isolates soon become necrotic ring-spots		*Gomphrena globosa* Local whitish or fawn necrotic spots or rings after 4–7 days, later developing reddish margins
Potato virus Y	Local chlorotic spots after 4–7 days, sometimes absent	Systemic fine vein clearing of young leaves after 8–14 days, later becoming a dark vein banding with mild mottle. Severe strains may cause stunting and necrosis	Faint local chlorotic spots after 5–8 days. Local leaves soon become necrotic	Severe systemic vein clearing, mottle and distortion. The whole plant is often killed-especially in winter	Numerous local whitish-green dots after 5–7 days, enlarging slightly		*Gomphrena globosa* No local lesions; some strains produce systemic pale green mottle, with semi-necrotic flecks

Potato virus Y (veinal necrosis or Y^N strain)	Local chlorotic spots after about a week	Systemic vein clearing on younger leaves, rapidly developing to a severe veinal necrosis with consequent leaf buckling and characteristic terminal down-curling or pursing of younger leaves		Severe systemic mottle, leaf distortion and usually, fairly rapid death of the plant	No infection	
Tobacco rattle disease virus	Brownish local necrotic spots after 4–7 days	Some isolates become systemic, causing necrotic flecks, veinal necrosis, or ring patterns, often with considerable leaf distortion	Local chlorotic or necrotic spots after 4–7 days	Some isolates produce systemic necrotic flecks, leaf buckling and distortion	Local chlorotic or semi-necrotic spots after about a week; lesions usually have a central necrotic speck	*Phaseolus vulgaris* (cv. The Prince) Tiny local necrotic dots after 1–3 days; lesions very small in summer months

	Nicotiana tabacum (cv. White Burley)		*Nicotina glutinosa*		*Chenopodium amaranticolor*		Other test plants
	Local	Systemic	Local	Systemic	Local	Systemic	
Tomatoes							
Cucumber mosaic virus	Local chlorotic (or necrotic) spots after 3–7 days	Systemic vein clearing and distortion of very variable colour and intensity with different strains; severe strains lethal in winter, mild strains may show very slight mottle on one young leaf, followed by symptomless young growth	Similar variable mottle and leaf distortion. Usually, a given strain causes more severe symptoms on this species than on tobacco		Numerous local pale green or whitish dots after 3–6 days		*Cucumis sativus* (cv. Butcher's Disease Resister). Different strains cause variable mottle, leaf distortion and stunting
Tomato aspermy virus	Local chlorotic spots after 4–7 days (not always present)	Systemic vein clearing, mottle, often with necrotic speckling on some leaves, and usually moderate or severe leaf distortion and blistering	Diffuse local chlorotic spots after 5–8 days	Systemic mottle, stunting, with severe and characteristic leaf distortion; young leaves often deformed to tendril shape. Usually kills the plant within a month in winter	Local yellow or pale green dots after 3–6 days, remaining discrete		*Cucumis sativus* (cv. Butcher's Disease Resister). Local chlorotic dots or small spots on inoculated cotyledons after 6–12 days. The virus does not become systemic, nor do true leaves show any reaction if inoculated with aspermy virus

Tomato spotted wilt virus	Local lesions variable; usually necrotic spots, rings or irregular local lesions after 3–7 days, usually enlarging considerably and sometimes leading to total necrosis of inoculated leaves	Many isolates become systemic, causing necrotic flecks, rings or veinal necrosis which may kill the plant	Large, local necrotic spots after 4–9 days; sometimes local lesions are initially small but soon enlarge	Most isolates become systemic, with necrotic flecks and streaks, often along the small veins. Necrotic areas may develop; apical necrosis may kill the plant	Local necrotic dots or ringspots after 3–7 days	
Potato virus X	Variable local chlorotic or semi-necrotic spots and rings after 5–8 days; symptoms vary greatly	Mild strains may cause only slight systemic spot-mottle followed by vein banding of the young leaves. Severe strains may induce considerable necrosis, severe mottle, necrotic veinal flecks, distortion and stunting	Local chlorotic spots	Variable systemic speckle-mottle or more severe distorting mottle and necrosis	Numerous pale green or fawn, local necrotic dots after 5–7 days; lesions from some isolates soon became necrotic ring spots	*Gomphrena globosa* Local whitish or fawn necrotic spots or rings after 4–7 days, later developing reddish margins

	Nicotiana tabacum (cv. White Burley)		*Nicotina glutinosa*		*Chenopodium amaranticolor*		Other test plants
	Local	Systemic	Local	Systemic	Local	Systemic	
Tobacco mosaic virus	Most *tomato* isolates produce chlorotic or necrotic local spots after 3–6 days. Most *tobacco* isolates show no local lesions	*Tobacco isolates* become systemic, with a rather slowly-developing mottle. Severity of symptoms varies greatly, from slight green mottle with mild distortion, to severe leaf distortion and narrowing, dark green blister mottle, or necrosis. Symptoms of some strains can be confused with those of aspermy on tobacco	Local grey or brown sunken dots or small spots after 2, occasionally 3, (some isolates consistently 4) days; lesions enlarge, and usually develop darker brown rims		Local chlorotic dots and flecks after 5–6 days	Some, but not all *tomato isolates* become systemic, causing yellowish flecks and spots, leaf buckling and *apical stunting*	

Tomato black-ring virus	Local semi-necrotic rings and chlorotic spots after 4–6 days often becoming necrotic whitish rings	Systemic chlorotic and necrotic ring and line patterns and concentric rings. With mild strains (such as the tomato or 'type' strain) symptomless young leaves are soon produced	Tomato strain usually symptom-less or nearly so		Faint local chlorotic or semi-necrotic dots and flecks after 3–7 days	Systemic apical leaf curling, flecking and often necrosis follow, varying with season	*Cucumis sativus* (cv. Butcher's Disease Resister). Mottle, followed by some weeks later by crater-like enations of a normal green on the under-side of the younger leaves
Tomato bushy stunt virus	Local reddish brown spots after 4–7 days, later enlarging slightly		Local necrotic brown spots after 3–5 days enlarging to 5 mm diam. with dark brown margins		No infection		*Gomphrena globosa* Local necrotic brown spots and rings after 10–14 days

Graft Transmission

One of the first demonstrations of the character of plant viruses by early investigators was the transfer of virus from infected to healthy plant when the two were grafted together. For many years such transfer was the only means of demonstrating the presence of a transmissible agent and, for many virus–host combinations, this still remains a valuable test. Viruses which become systemic in their host can usually be graft transmitted providing the donor and recipient host are compatible. In cases where graft transmission has been used as a diagnostic test, the recipient plant is chosen because virus presence induces symptom production; it therefore acts as an indicator plant.

Techniques for graft transmission are concerned with the art of grafting, the transfer of virus then being a consequent natural result [3]. Thus, the techniques vary according to the type of tissue in use. For the sake of convenience techniques are grouped into those for herbaceous subjects, tuber grafting and techniques for woody hosts. In all cases the graft will be successful if the meristematic cells of the vascular tissue of donor and recipient are brought together.

Techniques for herbaceous subjects

Materials

1 Sharp scalpel.
2 'Stericrepe' (Boots the Chemists).
3 Glass tubes 20×60 mm.
4 Sticks for supporting taller plants.
5 Test and donor plants.

The cleft graft

'Donor' plant must be actively growing.

1 Chose 'recipient' or 'scion' shoot of approximately similar diameter to that of 'donor' or 'stock'.

2 Remove the top part of the stock plant by making two downward oblique cuts.

3 Remove the top of the scion or indicator plant making similar downward cuts at a point where the stem diameter is similar to that of the stock.

4 Insert scion shoot into stock stem so that they fit well.
5 Bind the union firmly with 'Stericrepe'.
6 Support the scion with a stick if necessary.

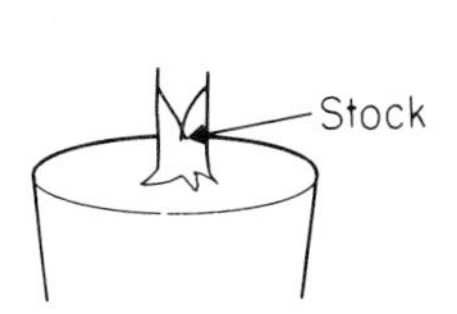

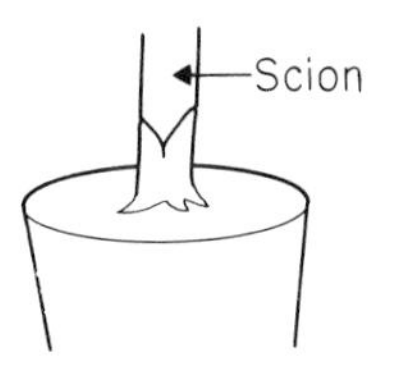

The approach or inarch graft

In this technique both stock and scion are maintained on their own root system. As before try to chose plants with similar stem thicknesses.
1 Make a downward oblique cut into stock plant stem at least 150 cm above soil level, and to half the stem thickness.
2 Make an upward oblique cut into scion stem at the same height and to the same depth.
3 Insert scion into stock, bringing the two stems together and taking care not to break either.
4 Bind with Stericrepe and support with a cane as before.

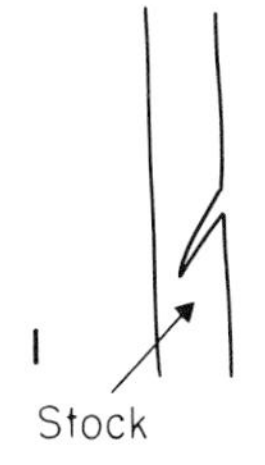

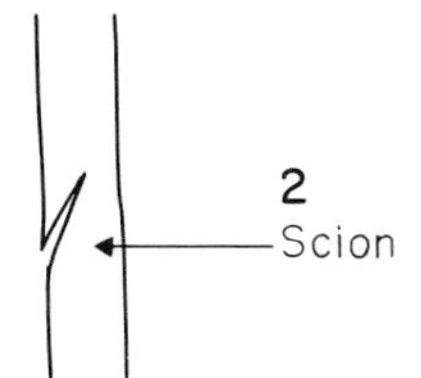

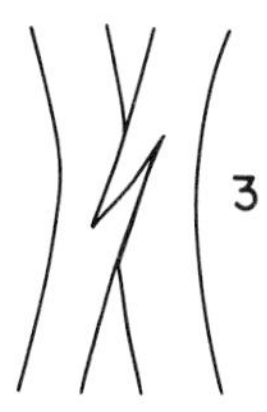

The bottle graft

A modification of the inarch graft can be used when either scion or stock (but not both) has been cut from its root, e.g. testing for chrysanthemum stunt viroid. The procedure is the same, except that the lower end of the plant without root is provided with water in a small tube or bottle to keep it alive whilst the graft takes and virus (or viroid) transfer takes place. The more usual procedure is for the indicator plant to retain its root system whilst the donor plant is a cut stem.

Tuber grafting

The most frequently used technique for grafting tissue from one tuber to another is that of core or plug grafting.

Materials

1 Virus-infected tuber.

2 Healthy tuber of approximately the same size as the infected tuber (cvs of each chosen according to requirement of test).

3 Cork borers size 13 and 13·5 mm.

4 Small dish with low melting point (42–45°C) paraffin wax.

Procedure (see Fig. 4.6)

1 Remove a plug of tissue from a flat area of the virus-infected tuber using the smaller cork borer. This plug should be deep enough to pass through the vascular tissue and a little beyond and should *not* contain an eye.

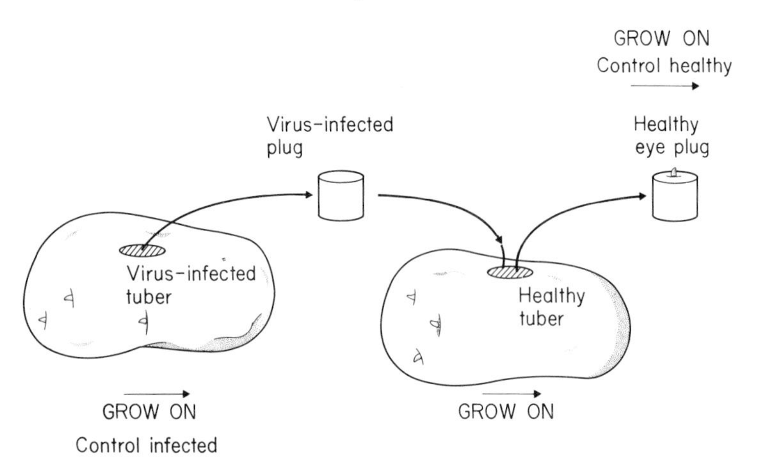

Fig. 4.6. Tuber grafting.

2 Remove a similar plug from the healthy tuber, this time *with* an eye, using the larger cork borer; it should be a little longer than the infected plug but should contain vascular tissue.

3 Insert the virus-containing plug into the hole in the healthy tuber. Take care not to damage the tuber skin.

4 Immerse the grafted tuber portion in melted wax so as to cover the cut surfaces to prevent desiccation and rotting.

5 Plant tuber and observe.

6 The virus-infected tuber and healthy eye plug may be retained and grown-on as a control.

Techniques for woody hosts

A variey of techniques are used for graft-transmission testing of woody hosts. Invariably, combinations of scion wood or bud material suspected of containing virus are grafted onto a specific growing rootstock along with indicator wood or bud. Recent modifications of the methods for graft-transmission testing for top fruit viruses involve the use of very small seedling rootstocks in pots, under glasshouse conditions, which can be exposed to artificial seasons thus speeding up the test. The grafting techniques required for each kind of test are the same, simply differing in scale. However, with traditional graft-transmission testing on outdoor rootstocks it is essential to carry out the particular technique at the appropriate time of year.

Double budding, needed for the detection of most tree fruit viruses, is done in the summer months, particularly July and early August. Grafting is usually done in the period from February to April. Rootstocks should be spaced 30 cm apart with 1 m between the rows. A gap, wide enough for a tractor, should be left either between each row or at least after every fourth row, to enable easy routine spraying and weed control to be carried out.

Materials

1 Budding knife.
2 Grafting knife.
3 Sharpening stone ('oil' or 'water').
4 Safety razor blades (optional).
5 Small hand-saw.
6 Secateurs.
7 Cold grafting wax (proprietary).
8 Grafting bandage.
9 Vaseline petroleum jelly.
10 Raffia.

Double budding

This is the technique most commonly used for virus testing. Two buds are introduced into the rootstock at an interval of about 35–50 cm. The top bud must be in a line directly above the bottom bud. The upper bud is from a known healthy indicator cultivar and the lower one from the tree to be tested for virus. Both scion buds are 'shield' budded into the rootstock and secured with raffia.

Procedure

1 Prepare buds from shoots of the current season's growth as follows. Hold the bud stick by the proximal end and remove the first good bud nearest to this by a shallow, slicing cut which begins 10–20 cm below the bud, passes under it and comes out well above the bud (see Fig. 4.7a). If the knife blade is arrested

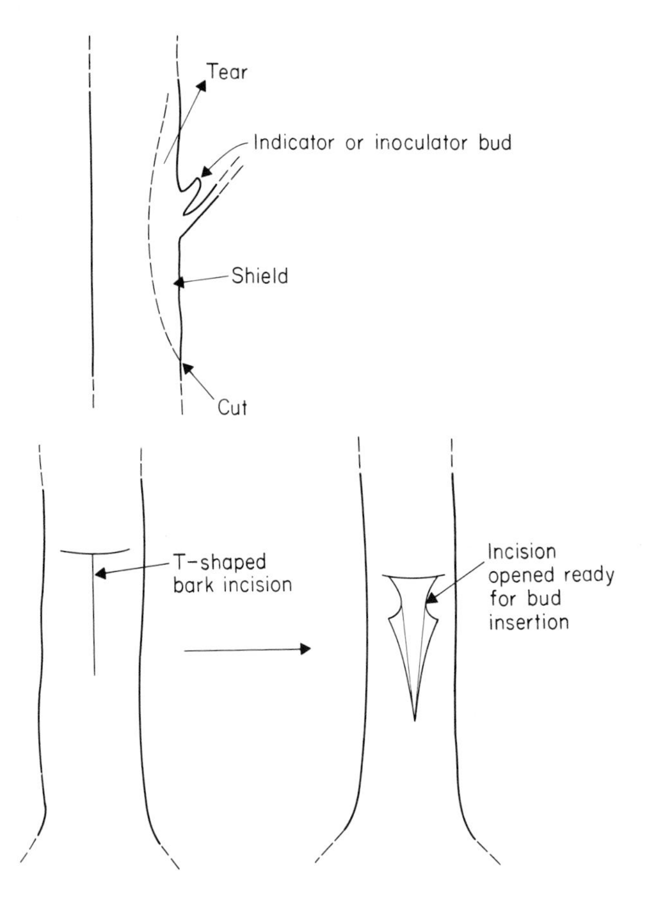

Fig. 4.7. (a) Removal of indicator or inoculator bud. (b) T-incision on virus-free rootstock needed for bud insertion.

just before it reaches the surface at the finish of the cut, the removal can be completed by tearing, which leaves a convenient strip of bark for use as a handle when inserting into the cut on the rootstock.

2 Prepare the rootstock by making a T-shaped incision in the bark (see Fig. 4.7b).

3 Raise the bark and insert the bud by sliding the shield downwards under the lifted bark so that the bud lies between the edges of the bark and well below the horizontal incision.

4 Cut off the 'handle' on the bud exactly at the horizontal incision and tie firmly with raffia (see Fig. 4.8).

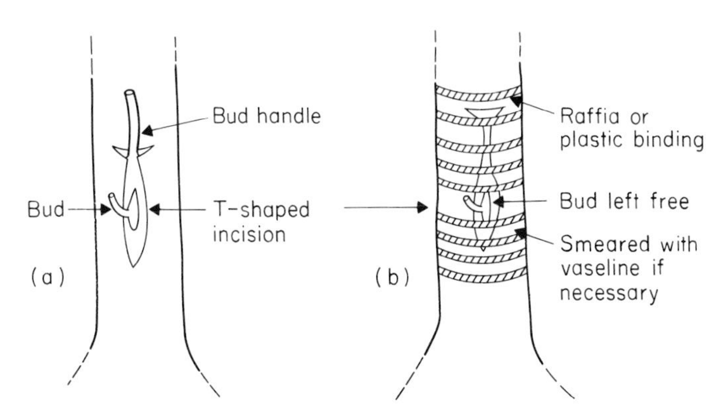

Fig. 4.8. Bud insertion (a) and tying (b).

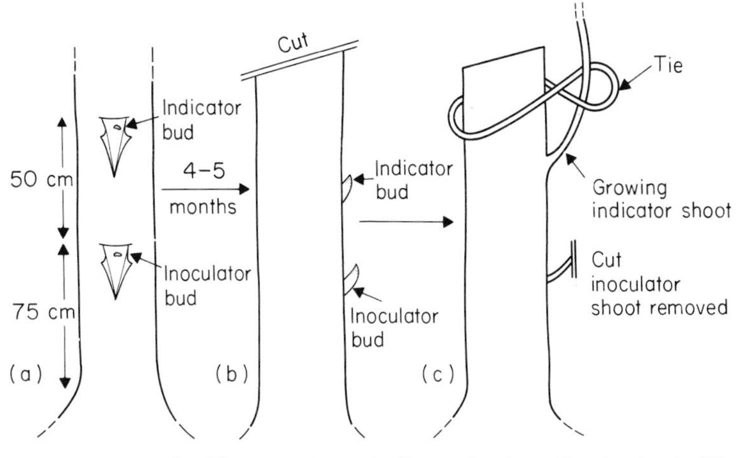

Fig. 4.9 Treatment of budding test plants. (a) Front view immediately after budding. (b) Side view in the winter after budding. (c) Side view in the spring after budding.

Bud sticks can be stored in a refrigerator at 4°C for several weeks if necessary.

In the winter after budding the main stem of the rootstock should be cut approximately 230 cm above the indicator bud. The cut surface should be treated with soft grafting wax. In the spring after budding the shoots produced by the inoculator and indicator buds will have grown to approximately 80 cm in length. At this stage the lower shoot is severed close to the rootstock. The indicator shoot is then grown on and initially tied to the butt of the rootstock for support (see Fig. 4.9).

Double grafting

This method involves incorporating a short piece (approximately 80 cm) of inoculator shoot between the virus-free rootstock and the indicator shoot.

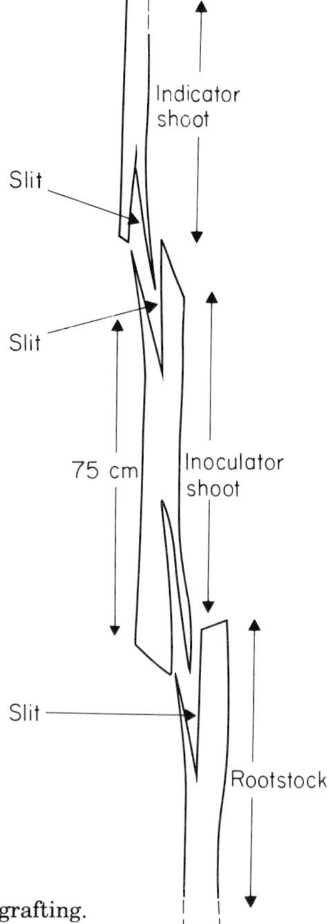

Fig. 4.10. Whip and tongue grafting.

Two whip and tongue grafts (Fig. 4.10) are used, one between the stock and inoculator, the other between the indicator and inoculator. Young bud-wood is used for both inoculator and indicator shoots.

Procedure

1 Make an oblique starting cut, approximately the same width as the thickness of the inoculator scion, at the distal end of the rootstock.

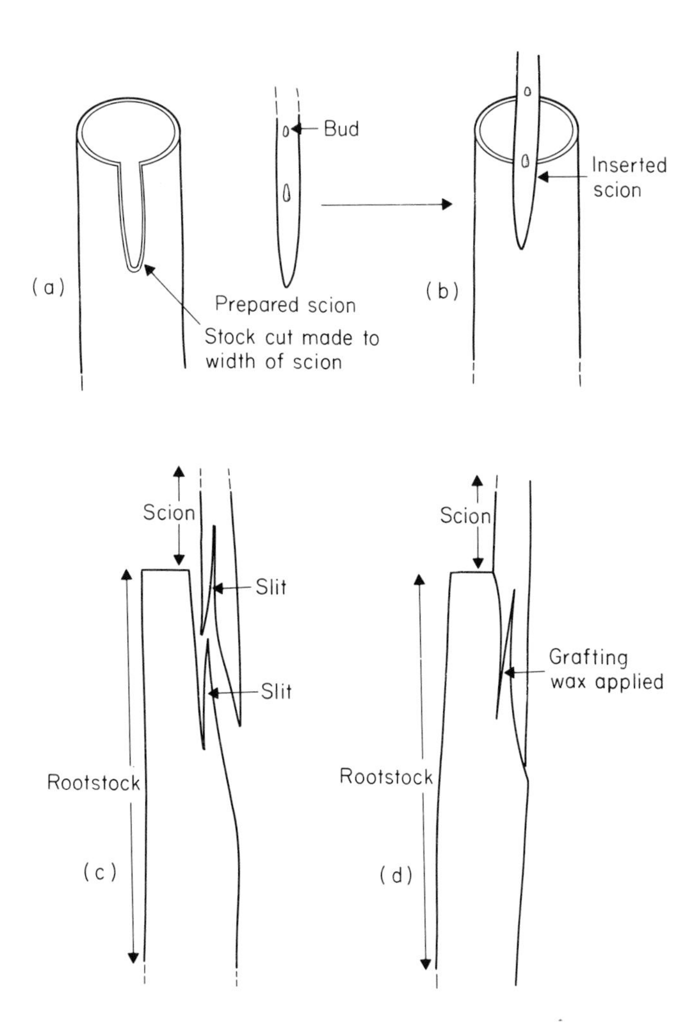

Fig. 4.11. Whip and tongue graft between inoculator scion and rootstock of different diameters. Views of the prepared stock and scion wood from front (a) before and (b) after insertion, and from side (c) before and (d) after incision.

2 Cut a downward-pointing tongue in the apical half of this slanting surface.
3 Similar corresponding cuts are made in the upper and lower half of the inoculator scion, and indicator scion.
4 Interlock all three pieces by their tongues, ensuring that cambial regions are in contact over as great a length as possible (see Fig. 4.10).
5 Tie and seal all cut surfaces with grafting wax.

If the rootstock is considerably larger than the inoculator scion, it may only be possible to match the two on one side, as both cambial regions have to be in contact (see Fig. 4.11). The indicator scion is allowed to grow on to develop symptoms.

Vector Transmission

For viruses which are not sap transmissible, transfer from suspect to indicator host using the natural vector may be the only means of detecting virus presence. Until recent improvements in virus purification techniques made antiserum production and serodiagnosis possible, vector transmission was the only diagnostic test for viruses of the luteovirus group. Within this group, strains of virus may be distinguished on the basis of their specific relationship with their vector and such strain differentiation remains a valid reason for use of vector transmission tests. An additional advantage of vector transmission tests is that they indicate the potential for spread of the virus from the host material, i.e. the ability of the vector to feed and transmit virus from host material of the kind under test, and thus have value in broader epidemiological studies. Feeding of trapped vectors on test plants enables determination of infectivity of vector populations.

If natural virus vectors are to be used in testing, it is necessary to understand the various ways in which the virus is acquired and transmitted by its vector. Insect and nematode transmitted viruses can be separated into three distinct groups, according to the manner in which they are transmitted: non-persistent or stylet-borne, semi-persistent, and persistent or circulative. These distinctions are particularly relevant to the mechanics of vector transmission testing.

Non-persistent viruses are acquired after only a few minutes feeding and are carried by the aphid for only a few (usually less than 4) hours at about 20°C.

Semi-persistent viruses are acquired most effectively after several hours feeding and persist in the aphid for 10–100 hours.

Persistent viruses require a long (24 hours) acquisition feed and sometimes a latent period before successful transmission. They persist in the vector for the remainder of its life.

Aphid-borne viruses may be transmitted by any one of the three methods. Leafhoppers, treehoppers, planthoppers, whiteflies and thrips usually transmit viruses in the persistent manner, although in some cases the mode of transmission may be more accurately described as semi-persistent. For nematodes, acquisition periods are difficult to define and although acquired virus is retained often for several weeks, it may be lost as the nematode moults. Beetle-vectored viruses usually occur in high concentration in their host, and transmission is thought to be due to regurgitation of sap containing virus during feeding and approximates most nearly to a semi-persistent mode.

Clearly, since the objective in vector transmission testing is to transfer virus from suspect to indicator plant successfully, the best conditions for vector feeding to encourage virus acquisition must be provided. Thus, whilst in general respects similar, there are important differences in technique when testing for viruses which may be transmitted in the non-persistent, semi-persistent or persistent manner. Some knowledge of the likely virus or viruses present in infected material is therefore necessary before vector transmission tests are attempted. Additionally, it is important to appreciate that some viruses may be confined to particular plant tissue and that vector feeding from such tissue is essential for effective transmission.

Aphid transmission

Techniques for handling different insect vectors vary, but certain principles are basic. The two broad techniques described below are for tests using aphid vectors.

Transmission of non-persistent viruses

Materials

1 Vector aphids (see p. 78) (must be aviruliferous).

2 Petri dish.

3 'Cling film' to close dish.

4 Indicator plant species (young, vigorously-growing specimens are best).

5 Insecticide and sprayer.
6 Small camel hair paint brush.
7 Small piece of filter paper.

Procedure

1 Disturb the colony of aviruliferous aphids so that they withdraw their stylets. This can be done by carefully tapping the abdomen with the brush, or by breathing on them.
2 Carefully pick up individual wingless aphids (sufficient to provide at least five per indicator plant), using the tip of the moistened paint brush and transfer to the Petri dish lined with moist (not wet) filter paper.
3 Close the dish with the 'cling film'.
4 Store aphids in a cool shaded place for about 1 hour to starve them.
5 Open the 'cling film' by cutting a trap door.
6 Transfer aphids to infected plant material (detached leaves in a separate Petri dish if possible) to feed for about 2 minutes. Watch to see if aphids feed (see Fig. 4.12).
7 Transfer at least five aphids to each separate indicator plant as in (**1**) and (**2**) above.

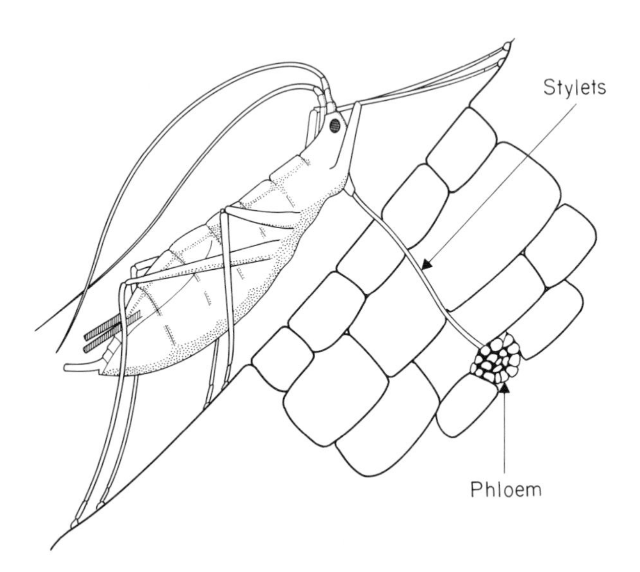

Fig. 4.12. Feeding aphid.

8 Cover to contain aphids and allow a transmission feed of no more than 1 hour (this may be done for small plants using inverted plastic beakers, the bottoms of which have been replaced by aphid-proof gauze).

9 Kill vector aphids by spraying or dipping whole indicator plant in aphicide solution and set test plant aside in glasshouse to grow-on for symptom observation.

N.B. Where mixtures may contain mechanically transmissible viruses to be left behind, avoid mechanical transfer by depositing aphids onto small pieces of clean filter paper on the indicator leaf surface. The paper can be discarded after aphids have moved onto and colonized the leaf.

For semi-persistent viruses increase the acquisition feed to 4 hours, then transfer the aphids to the test plants for a transmission feed of about 1 hour (but see p. 78).

Transmission of persistent viruses

Aphids are handled in the same way as for non-persistent virus transfer, except that no starvation period is required before acquisition and the transfer or inoculation feed must be longer.

Materials

As for non-persistent transmission.

Procedure

1 Transfer aphids to infected plant material in a Petri dish as in (**6**) above.

2 Leave in suitable cool shaded situation to feed for at least 24 hours.

3 Carefully transfer to indicator plants and leave for a further 24 hours. In aphid transmission of some viruses there is a latent period following virus acquisition before transmission can take place. Where such a period is known to exist, the duration of the transmission feed should be increased accordingly.

4 Kill aphids as before and set aside for symptom observation.

In all vector transmission tests, appropriate control tests must be made in order to ensure that the symptoms which develop are not simply due to feeding of aviruliferous vectors or extraneous factors, although the number of aphids employed in a transmission test should never be sufficient to cause damage in their own right. In a parallel control test, the appropriate vector should be fed on non-infected material and transferred to indicator plants using the relevant acquisition and transmission feeding times.

Whilst the acquisition feed may usually (although not invariably) be on detached leaves within a Petri dish or similar container, the transmission feed must be to plants which must then be grown-on to observe symptom development. A variety of insect 'cages' have been devised in which to achieve such transmission feeding. These vary from large insect-proof enclosures within which whole developed plants can be

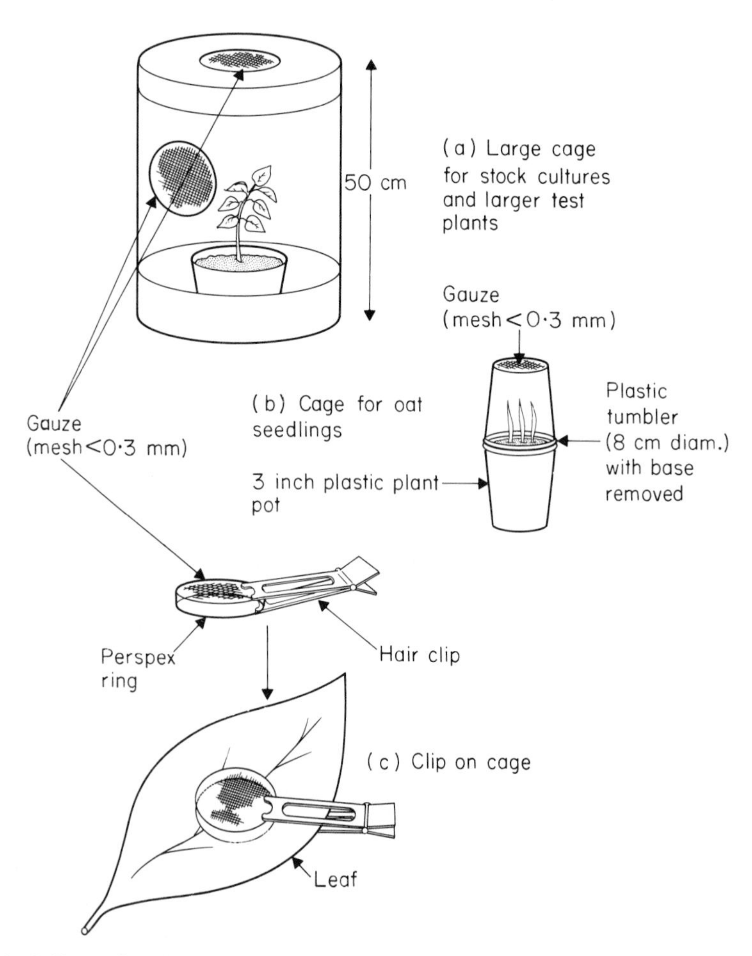

Fig. 4.13. Examples of insect 'cages'.

accommodated with the aphid vectors, to small cages which clip onto a small area of leaf and suitably confine vectors. A range of such 'cages' either specifically manufactured for the purpose or commercially available are illustrated in Fig. 4.13.

Development of symptoms on test plants

As with sap transmission testing, a vital component of the vector transmission test is the growing-on of the test plant after virus inoculation. Aphid-borne viruses become systemic, so that there is no need for marking of inoculated leaf. However, the speed with which symptoms become apparent may vary considerably. For the persistent viruses, e.g. barley yellow dwarf or potato leaf roll, symptoms usually do not show for 2–3 weeks, and test plants should be grown-on for at least 6 weeks to be sure no reaction will occur. Light and temperature conditions may affect the rapidity with which symptoms develop and the extent to which they are apparent. For luteoviruses, cool light conditions are essential, symptoms of some non-persistent viruses may be lost at higher temperature.

In using insect vectors for transmission testing for diagnostic or investigative purposes, it is vital to recognize that the success of the process depends upon providing the correct conditions for virus acquisition and transfer. Thus, the factors which govern the efficiency of insect transmission are not confined to test plant or *in vitro* virus preservation but must take account of the biology of the vector. The specific procedures described above are those used for aphid transmission testing. The principles behind these are appropriate to transmission testing using other insect vector types, although each insect vector has specific environmental requirements. It is not possible to cover each insect vector type in detail and for those other than aphids appropriate reviews should be consulted [8, 10, 14].

Selection of aphid species

The transmission of some viruses is limited to particular species of aphids and is unsuccessful using other species. This character frequently occurs, particularly amongst groups of viruses which are persistently or semi-persistently transmitted, and must obviously be recognized in diagnostic or experimental work. Within the luteovirus group, strains of some viruses—notably of barley yellow dwarf—have been separated using a combination of their vector specificity and the severity of symptoms they produce; only recently have these strains been confirmed by serological comparisons (a specific test procedure for BYDV testing is included in the Appendix, p. 88). Thus, it is vital to have some knowledge of the behaviour of the virus under investigation before choice of vector is made. Equally, it is important to be able to identify the vector species being used and distinguish it from contaminant vectors. For aphids, valuable information concerning vector/virus relationships may be found in reference [5] and identification keys [1] are available for cereal aphids.

Handling of aphids

The transfer of virus from one plant to another depends on the successful feeding by the vector aphid on both plants. Since the feeding apparatus of the aphid is complex and fragile, great care should be taken in handling them to avoid damage. Before transfer the aphid should be encouraged to withdraw its mouthparts (stylets) from the leaf on which it is feeding; movement of feeding aphids will inevitably result in damage. Feeding aphids should be disturbed on their food leaf by gently touching their antennae, or stimulated to cease feeding and withdraw stylets by breathing on them. Once the stylets are withdrawn the aphid can be carefully picked up with a paint brush, which may be used dry or slightly moistened but not wet. During this process avoid collecting surface wetness from the leaf; aphids will not transfer well if enveloped in a drop of water. At appropriate stages in the process of transferring aphids it is sensible to observe them using a hand lens. In particular, this can be valuable after transfer to ensure that they have survived the process and are feeding.

Effect of fasting on virus transmission

Non-persistent viruses are transmitted most efficiently when aphids are starved before the acquisition feed. The period of fasting or starving does not seem to be critical, periods of 15 minutes seem to be sufficient but longer periods may be more effective. With acquisition feeds of more than several minutes the effects of starving seem to diminish. Whilst it is possible that the prior starving of aphids might be expected to lead to avid feeding once food is available, it seems more likely that the effect may be related to the need for complete retraction of the stylets after one feed and before the next, a process taking several minutes. Starved aphids usually feed for a relatively short time, however, ending their first probe naturally within 1 or 2 minutes. These initial probes usually penetrate deeply into the plant tissue and result in the most effective virus transmission. Thus, the response to starving aphids appears to be a complex phenomenon.

A simplistic view of stylet-borne virus transmission would suggest that virus simply contaminates the stylet surface. Such, however, is far from true. Aphids which have acquired non-persistent viruses can transmit them repeatedly during the first hour after leaving the infected plants. The virus is therefore not a contaminant being wiped away by successive feeds (see also p. 78).

The increase in transmission efficiency achieved by starving aphids is not achieved with viruses which are persistently or semi-persistently transmitted.

Effect of temperature on transmission

Variation in temperature within the normal range at which aphids exist and multiply seems to have little effect on virus transmission. Perhaps the exception to this is found where persistently transmitted viruses have long latent periods before successful transmission feeding can take place. Experimentation has shown that at lower temperatures the latent period required is longer than at higher temperatures.

Condition and choice of test plants

In broad terms, the choice of indicator plants for insect transmission testing—as for sap transmission—relies on a knowledge of the virus or viruses involved, their host range and the reactions they produce. However, an additional consideration is that the insect vector should be able to feed on the plant. The range of indicator plants used is more specific for each virus than that for sap transmission testing and the appropriate literature must be consulted. Oat test plants, traditionally cv. Blenda but more recently the cv. Maris Tabard are good indicators of the presence of BYDV. *Physalis floridana* is used for detection of potato leaf roll, and *Senecio vulgaris* and *Capsella bursa-pastoris* for beet western yellows. Whichever plant is chosen it should be young and vigorous.

The need for 'helper' factors or viruses

There are examples of the need for 'helper' factors of some kind in virus groups which are transmitted non-persistently, semi-persistently and persistently. The complexity of the nature of these requirements is beyond the scope of this volume and the appropriate literature should be consulted [4]. However, a basic knowledge of the mechanisms which are involved is essential if experimental transmission of the viruses involved is envisaged.

Within the potyvirus group it has been shown that a component of cell sap must be acquired by the aphid before virus transmission can occur. Aphids fed on pure virus preparations through artificial membranes were unable to transmit virus. Additional evidence suggests that the helper factor is stimulated by virus presence since there is some specificity between the helper factor and virus strain.

The need for the presence of a helper virus for successful transmission of some semi-persistent and persistent viruses has been illustrated. For the semi-persistent ones this appears to depend upon the presence of particles of the helper virus within the aphid foregut, and in both the relationship seems to be very specific.

Culturing of aphid colonies

If aphid vectors are to be used regularly in transmission testing, cultures of relevant species must be maintained. The frequency of their use and the need for a range of species defines the sophistication of the culture system. Aphids grow and multiply well at temperatures between 15 and 20°C. Such temperatures, and the provision of continuous illumination suppress the development of winged aphids (alatae) which are not easy to handle and not such efficient vectors. Aphids prefer actively growing, healthy plants and care should be taken to ensure that the environment provided is suitable for plant growth.

Aphid cages

Only the larger of the types of cages used for housing test plants during transmission (Fig. 4.13) are large enough to accommodate stock aphid cultures. A variety of suitable structures can be made according to available resources. Traditionally, cages consisted of a wooden frame built over an enclosed tray. The top and sides were covered with aphid-

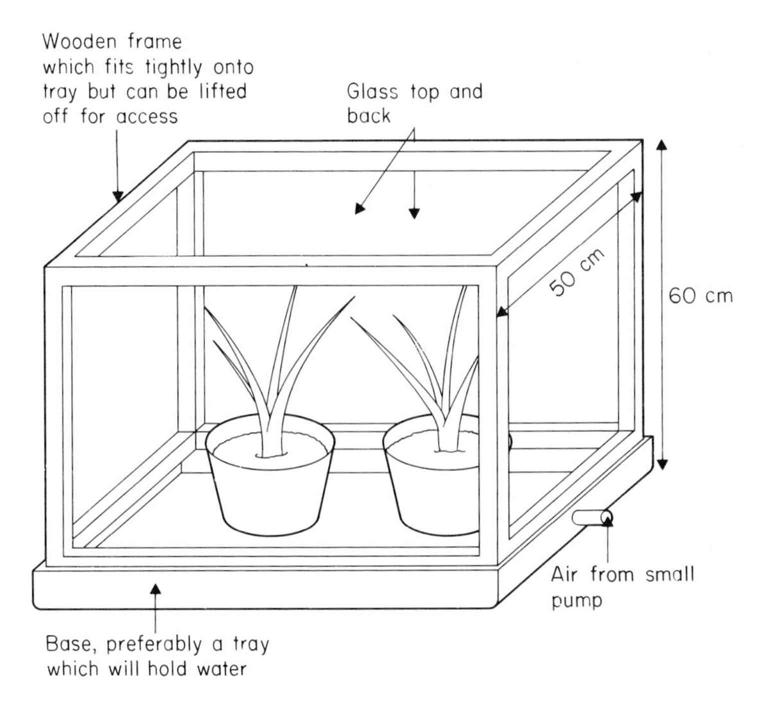

Fig. 4.14. Culture cage for aphid stocks. The sides and front are covered with gauze (mesh < 0·3 mm).

proof mesh, (meshes not wider than 0·3 mm), or some (top and back) surfaces were of glass or perspex to allow better access to light (Fig. 4.14). Such cages must be lifted off the plants for watering and plant replacement and this may allow aphids to escape. Additionally, wooden cages are more perishable and mould development may eventually accumulate on damp mesh. Single thickness mesh, whilst preventing aphid escape, does not prevent parasitization of aphids on the inner surface: Hymenopterous wasps on the outside may still attack aphids resting inside the mesh through the holes.

More suitable cages may be made wholly of Perspex with an integral floor and a door held in place by magnetic tape (Fig. 4.15). Such cages should have large ventilation holes which can be covered on the inner and outer surface with mesh, providing a double barrier through which parasitization cannot take place. Forced ventilation—by means of a separate pump provided with aphid-proof mesh in the back of the cage—is essential in Perspex structures to both reduce temperature and create a positive pressure within the cage, thus discouraging entry of other

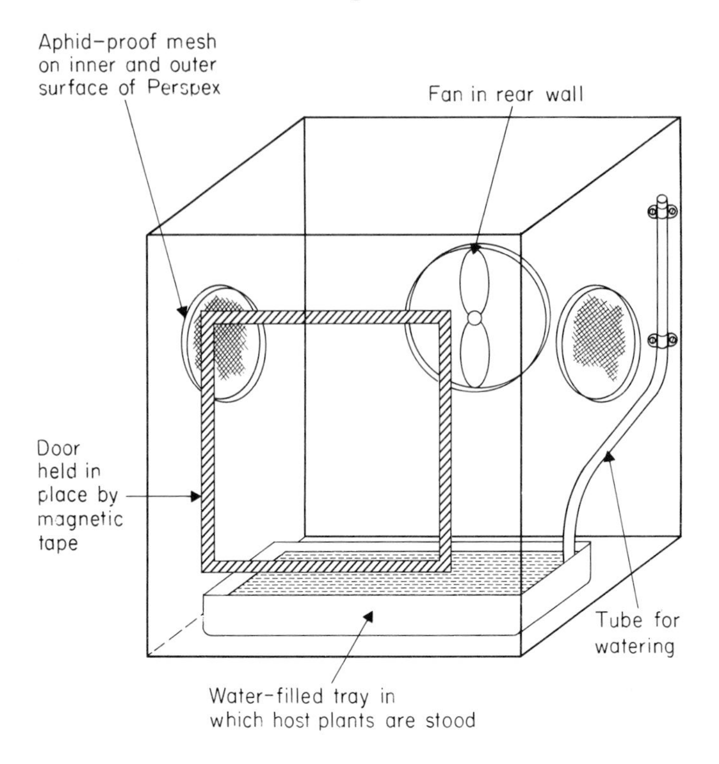

Fig. 4.15. Perspex aphid cage.

insects when the door is opened. It is vital to prevent condensation within cages which simply drowns aphids, but potted plants may be set in a water-filled tray which also serves to discourage itinerant aphids. Stock plants on which aphid cultures are maintained should be changed frequently (at least once per week) providing fresh young plants on which colonies will thrive.

Transfer of stock colonies can be achieved simply by tapping the leaves of the colonized plant sharply whilst holding the plant horizontally over an empty dry 28 × 16 cm sandwich box. Aphids fall into the box, and can then be tipped to one end and 'poured' onto the new plant.

It is essential to avoid acquisition of virus by stock cultures so that, if possible, plants for culture maintenance should not be hosts of the virus under study. For example, the maintenance of the aphid species *Myzus persicae* on chinese cabbage would be acceptable if they were to be used as vectors of potato leaf roll virus (which does not infect chinese cabbage), but not so if one of the brassica viruses was being transmitted. However, as might be expected, many aphid vector species only multiply readily on host plants to which they also specifically transmit viruses. In such cases regular tests of stock aphids are needed to ensure their freedom from virus.

Parasitization of aphid colonies by Hymenopterae which lay their eggs inside live aphid nymphs must be avoided. Regular check for parasitized aphids should be made. These are readily distinguished as swollen golden coloured individuals. If left within a stock the emergent insect will further colonize the aphids depleting the population. At first sight of such parasitization new colonies should be started in a clear area using eggs or very young aphids as starting material (see p. 79). Under suitable conditions aphids reproduce hermaphroditically and viviparously, so that starting from individuals should present no problem. In aphid stock maintenance, care should be taken to avoid other insect predators and fungal parasites, and where several species are concerned to avoid cross contamination.

Transmission of viruses by nematodes and fungi

Whilst a number of viruses are naturally transmitted by either nematodes or fungi, these organisms are rarely if ever used in diagnostic testing and infrequently in experimental situations. In part, this is because the range of viruses transmitted, particularly by fungi is limited, but probably mostly because the vectors are difficult to maintain in virus-free culture. However, the detection of viruliferous nematodes or fungi in cropped soils by bait testing, a form of vector transmission test is often useful [6, 10, 12, 13, 14].

Bait Testing for Virus in Soils

The need to know whether soil contains viruliferous vectors is often a prerequisite of planting new areas of perennial crops. The soil-borne vectors in such circumstances may be nematodes or fungi. Suitable samples of soil may be selected from the area to be planted and tested for the combination of virus and vector by bait testing. The procedure for detection of viruses with different nematode vectors is essentially the same as that for detecting those with fungal vectors. Bait testing may be done *in situ* in the field under test, but it is usually done under glasshouse conditions.

The following procedure has been found suitable for detection of nepoviruses, tobraviruses and some of those with fungal vectors (tobacco necrosis).

Procedure

1 Select soil samples from the field.
2 Whilst retaining soil in a moist stage, sieve through a coarse mesh to remove stones and larger debris.
3 If soil is excessively heavy, a little sterilized peat may be added. Such amendment should be kept to a minimum and no more than 50% of peat should be added. Dilution of field soil reduces the effectiveness of detection of virus, but soil must be in a suitable physical state to allow potting and seedling growth.
4 Moisten soil slightly if necessary but avoid overwatering.
5 Fill 5 inch plastic pots with soil under test. If possible ten such pots should be prepared.
6 Transplant bait seedlings, or sow seed, two or three per pot according to expected size, as appropriate for detection of the particular virus or group of viruses being sought (see p. 83).
7 Grow plants for 3–8 weeks under good glasshouse conditions ensuring adequate but not excessive watering.
8 In some bait plants, virus may become systemic during the growing-on period, producing recognizable symptoms. More frequently, the virus remains restricted within the roots. Unless clear systemic symptoms occur after the prescribed growth period, roots must be extracted and virus tested.
9 After growing-on, carefully tip out the pot contents and separate out the plant roots. Wash off all soil particles with two successive changes of water so that roots are clean. Take care to lose as little root as possible.
10 Pool roots of plants from pairs of pots and grind in an appropriate buffer with an abrasive for sap transmission.
11 Inoculate to individual test plants which give reactions to the virus being sought and observe symptom development.

Factors affecting success of detection

By their very nature soil-borne viruses, both nematode and fungally vectored have patchy distribution within a cropped area. Thus, sampling must be random if attempted in the absence of areas of diseased plants, and the chance of successfully detecting virus depends greatly upon sampling extent. As with most virus tests, therefore, the bait-testing procedure can only be used to demonstrate virus presence not its absence. Investigations should take account of all available information, e.g. records of 'poor' patches in previous crops, areas of different soil types which might particularly favour the virus vector, or the existence of herbaceous weed hosts of the virus.

Since in bait testing soils the living vector is used to transfer virus, care should be taken to maintain soil samples in a condition which preserves the vector in an active healthy state. Thus, extremes of temperature should be avoided and except in certain situations (see below) soils should not be allowed to dry out. Excessive vibration will cause the rapid death of some vector nematode populations.

Choice of bait plant

A variety of bait plants have been used and some are reported to be better for particular viruses than others. Fortunately, most nematode vectors feed on a wide range of plants so that convenience of management may be a factor in selection. For nepoviruses, turnip seedlings are most effective when planted into infective soil no more than 2 weeks after germination and grown on for 3 weeks. Cucumber may also be used for a variety of viruses but should either be sown directly into the soil being tested or transplanted before cotyledons separate. Germinating seeds of mung bean (*Phaseolus aureus*) in the soil is the best method of detecting tobacco necrosis virus transmitted by *Olpidium brassicae.*

Testing for potato mop top virus

Whilst the method described above is suitable for most viruses, bait testing for potato mop top virus requires fundamental changes in procedure [6]. Soil sampled from the area under test should be air dried at room temperature for 2 weeks, then remoistened and plastic pots filled. Seedlings of *Nicotiana debneyi* (as small as may be handled satisfactorily) should then be planted and grown-on under normal glasshouse conditions. After 5–6 weeks, when viruliferous *Spongospora subterranea* is present in the soil under test, characteristic necrotic markings should appear on the test plants as systemic virus invasion occurs. With this virus, washing out and testing of roots is unnecessary.

Bait leaf method for detection of tobraviruses

Tobraviruses may be detected by the basic bait plant method as described above. However, a modification using bait leaves may be quicker and more reliable [13].

Procedure

1 Prepare soil as before (sieve and moisten).
2 Fill plastic sandwich boxes (16×28·5×9 cm) with soil.
3 Bury leaves of *Nicotiana tabacum* 'White Burley' from soft lush glasshouse raised plants in the soil, four or five per box according to size.
4 Incubate boxes at 15°C for 3 days.
5 Remove leaves and rinse free of soil.
6 Lay leaves flat on damp filter paper.
7 Incubate at 20°C and 12 hours alternating light and dark for 3 days.
8 Grind leaves in buffer with abrasive and inoculate to *N. tabacum* 'White Burley' in the usual way.
9 Grow-on and record presence of local lesions after 8–10 days.

In all bait tests control pots may be prepared by autoclaving soil under test before proceeding with bait testing, but take care to sterilize thoroughly since some fungal resting spores are quite heat resistant and nematodes may encyst.

Seed Testing

Seed transmission of plant viruses is the exception rather than the rule. Where it does occur, in most host–virus combinations the proportion of seeds infected is usually small, but in some most seeds may carry virus. Not all viruses cause symptoms in their host when they infect through the seed. Thus, selection by visual inspection of apparent virus-free parent plants may often be an unreliable means of avoiding seed-borne virus spread. Similarly, the growing-on of large numbers of sample seed batches in the hope of detecting infected seedlings is likely to be ineffective. Specific tests may be needed therefore, to establish the absence of seed-borne virus, or that it does not exceed prescribed maximum tolerances. In all cases, an important component of the test procedure is the selection of samples and the statistical interpretation of results.

Perhaps the most basic method for detecting seed-borne virus consists of sap transmission testing the leaves of seedlings grown from seed under test. Such testing overcomes the problem of lack of seedling symptoms, but is demanding of space and is time consuming. For some viruses it is possible to sap inoculate extracts of the seeds themselves to indicator plants. In such a test the most reliable detection of lettuce mosaic virus in lettuce seed was obtained [7] as follows.

Procedure

1 Grind 300 seeds in a mortar with 5 ml 0·03 M phosphate buffer pH 7 containing 0·5% sodium acid sulphite and 0·5 sodium diethyldithiocarbamate.

2 Activated charcoal is added to the homogenate at the rate of 75 mg ml^{-1} to counteract inhibitors.

3 Rub homogenate paste onto leaves of *Chenopodium quinoa* plants previously dusted with carborundum. Plants should previously have been grown at 18°C with a 16-hour day.

4 Grow plants on at 25°C with a 16-hour day.

5 Observe development of virus symptoms (for lettuce mosaic virus a systemic mosaic develops).

Pregnated seed has also been used.

In many other virus–host combinations, serological methods have been used to detect virus. In particular, the more sensitive ELISA test with its ability to handle large numbers of samples has become widely used. Similarly, some viruses which reach high concentration in seeds can be identified in extracts examined by conventional quick electron microscopy. Considerable increase in sensitivity is gained when IEM techniques are used for this purpose.

Growing-on Test for Potato Tuber Viruses

Whilst not strictly a test for virus in the seed, the growing-on test, otherwise known as tuber indexing, which allows assessment of tuber-borne virus after harvest, may be of value. Such a test defines the actual virus level in a tuber stock, when no more virus spread can occur and seed tubers have been selected. Field inspections of parent crops may always be criticized since late season (post-inspection) virus spread may be undetected.

The accuracy of the growing-on test relies on the initial sample of tubers selected. Obviously this should be random, and if taken from the field after burning off haulm, should consist of separate single seed-size tubers from separate randomly-selected plants. Selection from stores after riddling may be more random but introduces undesirable delay. The procedure for the growing-on test is as follows (see Fig. 4.16).

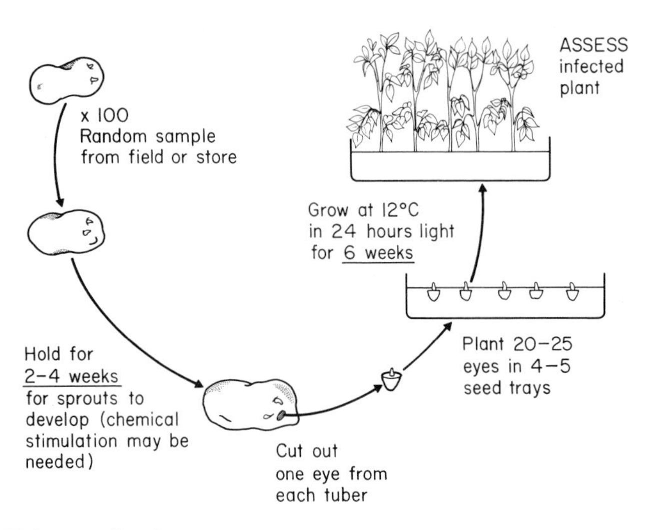

Fig. 4.16. Tuber test for virus.

Procedure

1 Select sample of tubers as described above (100 are often tested but the minimum virus level detectable is limited by sample size).

2 Store tubers at 20°C in light, humid conditions to promote sprout development. Chemical treatment may be needed to break tuber dormancy. The most effective treatment requires fumigation for 48 hours with a mixture of anhydrous ethylene chlorhydrin, dichloroethane and carbon tetrachloride (7:3:1 by volume) ('rindite'). This should be done in a proper fumigation chamber; inhalation of the volatile gases is harmful. Tubers may be treated at the rate of 1 ml 'rindite' per kilogram of tubers. Tubers to be treated should be clean and dry and should have mature skins. After fumigation tubers should be stored in a warm, light, humid environment when sprouting should begin in 2 weeks.

3 When most tubers have sprouts 2–3 mm in length an eyepiece (dormant shoot) should be cut from the rose (apical) end of each. A suitable tool for this purpose may be made from a domestic potato peeler. Care should be taken to ensure that the volume of tissue cut out is adequate to support the initial sprout growth. An 'eyepiece' should consist of a conical piece of tissue of at least 2 cm diameter and 2 cm depth.

4 Plant eyepieces at least 2 cm deep, regularly spaced in standard seed trays, 20–25 per seed tray in a compost which should not be provided with too much nitrogen fertilizer. A suitable compost may be made from one 350 litre bale of raw peat with the following added:

105 g ammonium nitrate;
525 g superphosphate;
350 g potassium sulphate;
1400 g calcium carbonate;
water to mix.

5 Grow-on in an aphid-proof glasshouse at 12°C with continuous supplementary light. Water regularly.

6 After 6 weeks growth, assess virus incidence on the basis of visual virus symptoms in the resultant plantlets. The percentage of plants infected by each virus can then be determined.

Unsatisfactory results will be obtained if plants are grown at higher temperatures or without adequate light.

References

1. Anon. *Key for the field identification of apterous and alate cereal aphids with photographic illustrations.* MAFF (Publications), Tolcarne Drive, Pinner, Middlesex HA5 2DT.
2. CMI/AAB. *Descriptions of Plant Viruses.* Commonwealth Agricultural Bureaux, Farnham Royal, Bucks.
3. Garner, R.J. (1958) *The Grafter's Handbook,* 2nd edn. Faber & Faber, London.
4. Govier, D.A. & Kassanis, B. (1974) Evidence that a component other than the virus particle is needed for transmission of potato virus Y. *Virology,* **57,** 285–286.
5. Harris, K.F. & Maramorosch K. (eds) (1977) *Aphids as Virus Vectors.* Academic Press, London.
6. Jones, R.A.C. & Harrison, B.D. (1968) Potato mop-top virus. *Annual Report of the Scottish Horticultural Research Institute, 1967,* pp. 59–60.

7. Marrou, J. & Messaien, C.M. (1967) The *Chenopodium quinoa* test: a critical method for detecting seed transmission of lettuce mosaic virus. *Proceedings of the International Seed Testing Association,* **32** (1), 49–57.
8. Noordam, D. (1973) Transmission of viruses (or mycoplasmas) by leafhoppers. In *Identification of Plant Viruses,* p. 159. Pudoc, Wageningen.
9. Raski, D.J. & Hewitt, W.B. (1967) Nematode transmission. In *Methods in Virology,* Vol. 1 (eds Maramorosch, K. & Koprowski, H.), p. 309. Academic Press, London.
10. Slykhuis, J.T. (1972) Transmission of plant viruses by eriophyid mites. In *Principles and Techniques in Plant Virology* (eds Kado, C.I. & Agrawal, H.O.), p. 204. Van Nostrand Reinhold, New York.
11. Taylor, C.E. (1972) Transmission of viruses by nematodes. In *Principles and Techniques in Plant Virology* (eds Kado, C.I. & Agrawal, H.O.), p. 226. Van Nostrand Reinhold, New York.
12. Teakle, D.S. (1972) Transmission of plant viruses by fungi. In *Principles and Techniques in Plant Virology* (eds Kado, C.I. & Agrawal, H.O.), p. 248. Van Nostrand Reinhold, New York.
13. Van Hoof, H.A. (1976) The bait leaf method for determining soil infestation with tobacco rattle virus—transmitting tinchodorids. *Netherlands Journal of Plant Pathology,* **82,** 181–185.
14. Whitcomb, R.F. (1972) Transmission of viruses and mycoplasma by the auchenorrhynchous homoptera. In *Principles and Techniques in Plant Virology* (eds Kado, C.I. & Agrawal, H.O.), p. 168–203. Van Nostrand Reinhold, New York.

Appendix: *Sap Transmission*

Indicator plants

Age (days from sowing at 20°C) and proper stage of use of test plants (number and stage of maturity of leaves). Times to reach maturity vary according to season.

Species	Age (days)	Development (leaves)
Avena sativa	5–10	1 (3 cm high)
Brassica pekinensis	21–28	5
Chenopodium amaranticolor	50–60	6 mature
Chenopodium quinoa	50–60	6 mature
Cucumis sativus	10	cotyledons
Datura stramonium	35–56	1 or 2
Gomphrena globosa	70	2–4 pairs
Nicotiana glutinosa	35–45	3 or 4
Nicotiana tabacum	35	3 or 4
Petunia hybrida	35	3 or 4
Phaseolus vulgaris	10	2 primary
Physalis floridana	28	4 or 5
Pisum sativum	17	3 or 4
Tetragonia expansa	28	8
Vicia faba	14	1 mature
Vigna sinensis	10	2 primary

Buffer additives to overcome inhibitors

Thioglycollic acid: used at 0·1% in maceration buffer.
Sodium sulphite: used at 0·1–0·5% or 0.02 M in maceration buffer.
Sodium diethyl dithiocarbamate (DIECA): 0·01 M in maceration buffer.
Nicotine: use as a 1% aqueous solution which should be adjusted to pH 8 with N HCl.

Care—Nicotine is carcinogenic!

PVP (polyvinylpyrrolidone): use at 100 g $litre^{-1}$ in phosphate (0·01 M pH 7·0) maceration buffer.

Appendix: *Aphid Transmission*

Testing cereal specimens for barley yellow dwarf virus.

Indicator plants

Oats cv. Blenda (cv. Maris Tabard is a suitable alternative) sown four to a 4 inch pot of compost, 5 mm deep.

Plants are ready for use when 20–50 mm tall, usually 5–7 days after sowing.

Vectors

Rhopalosiphum padi and *Sitobion (Macrosiphum) avenae.*

Acquisition feed

1 From suspect sample select ten leaves on the basis of:
(a) yellowing possibly caused by BYDV;
(b) suitability for aphid survival.

2 Cut two 60 mm pieces from each leaf and place the separated pieces into each of two Petri dishes or damp filter paper arranging the leaves in a criss-cross manner to avoid close packing. (Label dishes with specimen no., aphid species and date.)

3 Carefully introduce about twenty aphids of *R. padi* into the first dish and a similar number of *S. avenae* into the second. Take care in handling aphids, avoid touching by using a paint brush to urge aphids onto test material.

4 Store the dishes in a cool, well lit place.
5 Allow an acquisition feed of 2 days, but move on after 1 day if deterioration of leaves makes it necessary.

Transmission feed

6 Cover one seedling with a glass specimen tube to prevent aphid colonization (control).
7 Move at least ten aphids onto each pot of indicator plants. (Label with specimen no., aphid species and date.)
8 Store in a well lit, cool place. Cover with cages (beakers) or stand well separated in a tray of water.
9 Allow an acquisition feed of 2 days.
10 Examine plants for degree of survival of aphids.
11 Use metasystox or similar insecticide to kill off aphids (outside the glasshouse). After spraying, the pots should be kept in a confined space for 1 hour.
12 Examine to ensure aphids are dead.

Observation of test

13 Keep pots in good light, but cool temperature. Watering should be regular to avoid leaf scorch and mildew.
14 Examine at 14, 21 and 28 days for:
(a) orange-brown discoloration;
(b) scorching of leaf edges;
(c) reduction in growth compared with control;
(d) chlorosis and 'water soaked' patches when held to light;
(e) leaf edge cutting—as if leaf margin has been snipped with scissors.

Maintenance of test aphids for BYDV test

Aphids will multiply readily under cool, light conditions (16°C). Stocks of (virus-free) aphids should be kept on young oat plants of about 75 mm in height in cages. Plants should be watered regularly by standing pots in a saucer of water. Keep species completely separate (different sides of same room). The aphids will need new plants periodically (weekly) but not all cages should be changed at one time.

5

Serological Techniques

The outer coat or capsid of plant viruses consists of protein subunits of a type and in an arrangement peculiar to each virus. The virus particle, therefore, because of its specific three-dimensional shape and its size, forms an ideal antigen which on injection into the blood system of a suitable animal stimulates the production of antibodies. Such antibodies, present in the serum fraction of the blood, react only with the stimulating virus or with very closely related viruses. The specificity of this antigen (plant virus)/antibody (antiserum) reaction can be used in a variety of *in-vitro* tests to demonstrate the presence of a plant virus and to determine its character. Many plant viruses can be purified with comparative ease and the production of specific antisera of a high quality has become a routine procedure.

Virus purification follows a number of basic steps. First, perhaps by inoculation to a range of indicators, the purity of the starting virus material is checked. This is then inoculated to a host plant in which it can multiply rapidly by systemic invasion, but which does not contain tannins, gums, latex or phenolic compounds which will interfere with purification. Having determined the time at which maximum virus content occurs, and which part of the propagating host plant is best to use, the material is harvested. Using a suitable buffer, provided if necessary with additives to prevent virus inactivation, the harvested virus-infected material is homogenized in a way which maximizes virus release. The resultant sap may be clarified by a variety of means: centrifugation, filtration, heating, freezing and the addition of organic solvents are some of the more frequently used methods. Usually the next step is to concentrate the virus extracted by centrifugation, but again a variety of other methods are available, e.g. precipitation with salts or alcohols. Alternate cycles of high and low speed centrifugation will eliminate most of the particulate matter larger than the virus. Finally rate-zonal density-gradient centrifugation should remove all contaminating materials, leaving a pure virus preparation. Full details of the many important procedures involved in virus purification should be consulted [6, 9].

Antisera may be produced by immunizing rabbits or other suitable animals using purified virus. Several injections are administered intravenously or intramuscularly or in combinations of both over a period of several days. Test bleeds to check the level of antibody are made, 1–3 weeks after the immunization. When the antibody level is satisfactory, several larger bleeds can be made without harming the immunized animal. Such blood samples are allowed to coagulate, then, after overnight storage in a refrigerator, the serum may be decanted and clarified by low speed centrifugation. Reviews of antiserum production are available [4, 12].

The specific component of antisera is made up of globulins, and the type and amount of these determines the quality of the antiserum. Three main classes of immunoglobulin (gamma globulin) are known, IgG, IgA and IgM [12]. Each is made up of heavy and light chains linked together by disulphide bridges. The simplest is IgG which exists as a Y-shaped molecule, of which the arms of the Y are the combining sites which are specific for each antiserum (see Fig. 5.1).

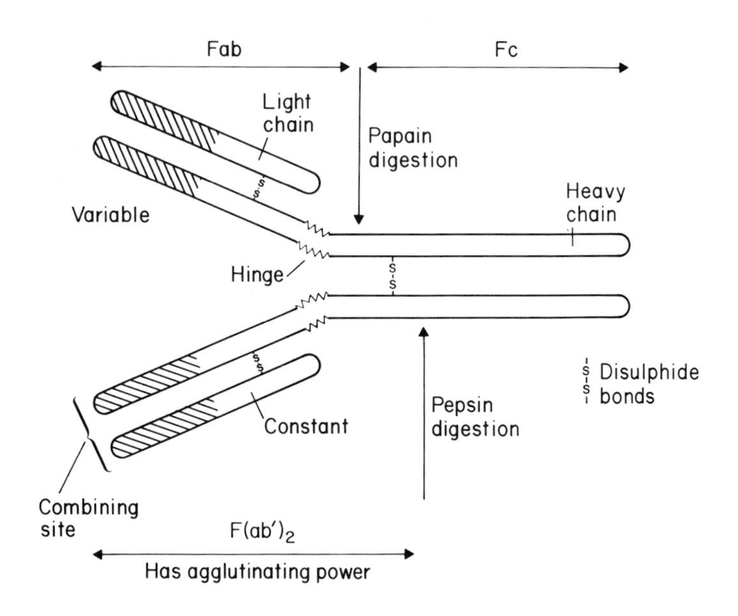

Fig. 5.1. Diagrammatic structure of the IgG molecule.

Serological reactions rely on the unique nature of the way in which the antigen and antibody molecules fit together. The combining sites on antibody molecules are complementary to different determinants on the surfaces of the homologous (stimulating) antigen. When antibody and antigen are mixed under suitable conditions, a visible precipitate results. Such a precipitate probably results from the linking of antigen molecules by antibody, forming an aggregate which precipitates when it is big enough. However, there is also evidence that the act of combination of the two components changes the physico-chemical state of the complex, rendering it insoluble.

Many *in-vitro* tests have been developed which use the specific relationship between an antigen and its (homologous) antiserum. Some of these serve mainly to establish the quality of the antiserum, others may be of great value in defining relationships between viruses and yet others serve as sensitive, specific diagnostic tests. It is the last group which will form the body of the techniques in this chapter. Further information about others can be found in reference [15].

The tests described can be conveniently divided into four different types: precipitation, agglutination, immunodiffusion and labelled antibody techniques. A fifth group of tests using antisera are those which involve immunosorbent electron microscopy but these will be discussed in Chapter 6. The first three types of test involve the visualization of antigen/antibody precipitates in one form or another, and it is important to understand the conditions under which such precipitates can form.

The representation of the antigen/antibody precipitation process as the formation of an insoluble lattice may be an oversimplification but it illustrates the need for careful balancing of the concentrations of the two components. Figure 5.2 illustrates the way in which excess of either component can prevent lattice formation and the build up of visible precipitates. With antigen excess (a), there are insufficient antibody molecules to link antigen molecules, and with antibody excess (b), all antigenic sites are filled and no linking occurs. It is essential, therefore, to use a range of dilutions of both antigen and antibody in order to ensure at least an approximation of the optimal proportion ratio (c).

A number of other fundamental principles should be appreciated before proceeding to serological testing. It is important to realize that in the preparation of pure virus suspensions for use in inoculations to stimulate antibody formation, contaminating plant sap proteins will also readily stimulate antibodies. Such anti-plant antibodies, albeit usually at low levels in antiserum preparations, may cause non virus-specific reactions. To avoid confusion it is essential to incorporate appropriate controls, using healthy sap extracts in addition to those suspected of containing virus. Similarly, in addition to antisera to plant virus, tests should always incorporate controls using normal serum (that present in the blood system of the animal prior to inoculation with antigen). Antisera are usually diluted for use in saline solutions and virus or sap extracts in buffered saline.

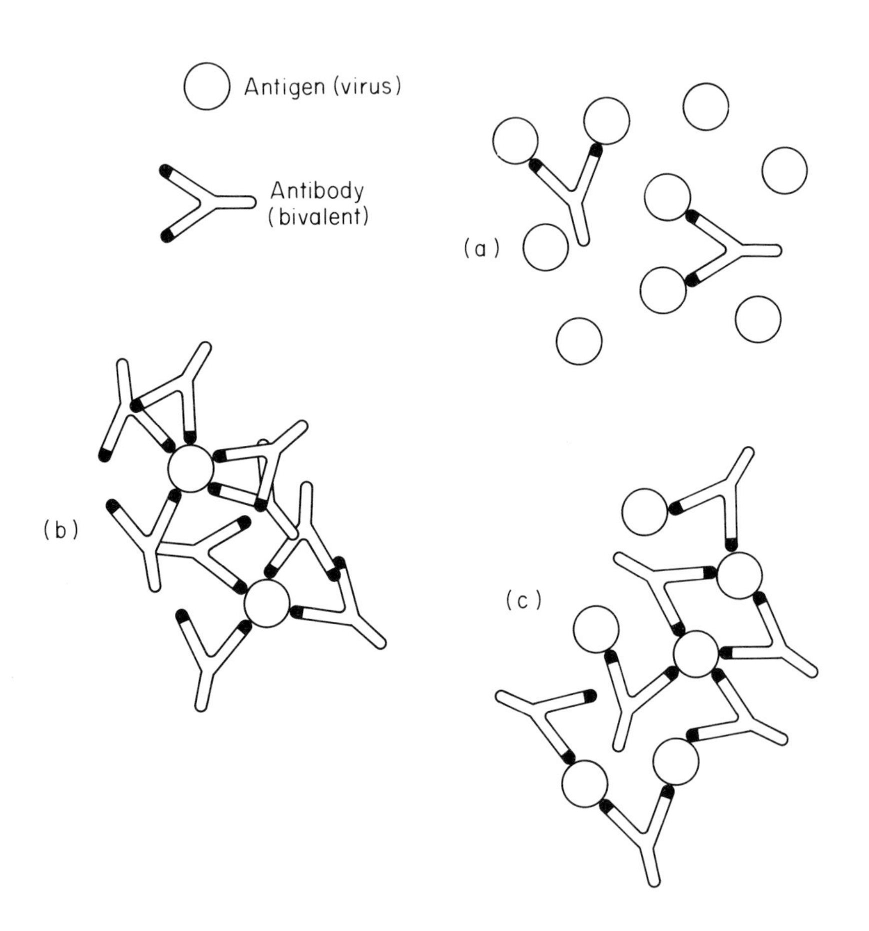

Fig. 5.2. The antigen/antibody precipitation process: (a) antigen excess and (b) antibody excess form soluble invisible precipitates; (c) antigen and antibody in optimum proportions gives a visible lattice. (After Cushing & Campbell (1957), McGraw-Hill.)

Precipitation Tests

Tube precipitin

Whilst infrequently used directly as a diagnostic test, a knowledge of the principles behind the tube precipitin test is important, since it is by this method that the titre of an antiserum can be determined for use in further serological investigations. The test depends on the formation of a visible precipitate in small test tubes when antigen and antibody are mixed in suitable conditions. Reactants are usually diluted in neutral 0·85% saline because a precipitate will only form in the presence of electrolytes and most viruses are stable at neutral pH.

Materials

1 Precipitin tubes (approx. 7×100 mm): these can be made from tubing of an appropriate bore (unconstricted small freeze dry ampoules make ideal precipitin tubes).
2 15 ml tubes in which to prepare virus or antiserum dilution.
3 Suitable racks to hold tubes.
4 Water bath at 37°C.
5 Pipettes, 1 and 5 ml for diluting antigens, antisera and adding saline—or adjustable hand pipettes with interchangeable tips—range up to 100 μl.
6 Saline (0·85%), or phosphate buffered saline (PBS) (see Appendix).
7 Light box with slit light source.
8 Antiserum and normal serum.
9 Clarified sap containing virus (clarify by centrifugation at 5000 **g** for 15 min).

When first using an antiserum/antigen combination, observation of the reactions between a range of antiserum and antigen dilutions as described below should give (a) the antiserum titre, i.e. the highest antiserum dilution giving a positive reaction, (b) the virus dilution end-point in the sap used, (c) the optimal proportion ratio. Ideally, virus of a known concentration should be used but in many cases this is not available. If possible, look up the properties of the particular virus in use (see CMI sheets), e.g. TMV occurs at high concentration in saps, and dilutions up to 1/4096 may be appropriate. However, for most viruses dilutions up to 1/128 or 1/256 are more suitable.

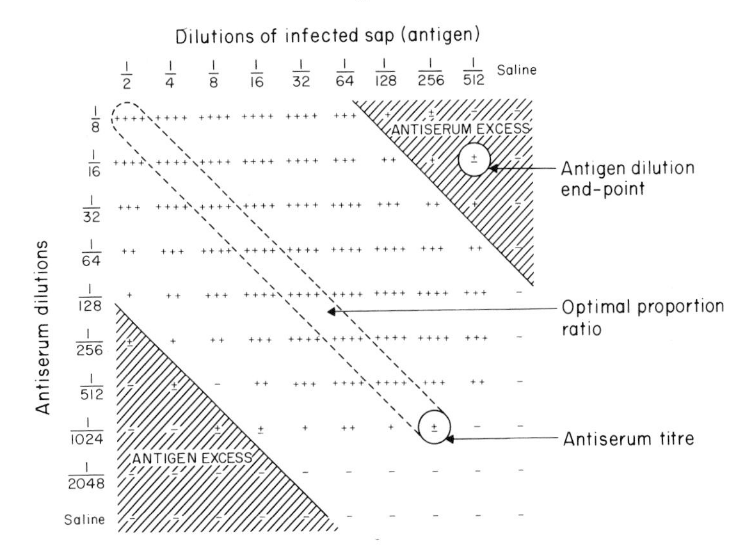

Fig. 5.3. Tube or microprecipitin test layout.

Procedure

1 Arrange the precipitin tubes in rows in racks according to the arrangement in Fig. 5.3.

2 In another test tube rack arrange two rows of nine large tubes and add 5 ml of saline to all eighteen tubes except the first in each row.

3 For antiserum dilutions add 1·25 ml of antiserum to the first tube and make up to 10 ml with saline. Mix thoroughly. Transfer 5 ml to the second tube, mix thoroughly and transfer 5 ml to the third tube etc., until eight twofold dilutions are made (1/8 to 1/2048).

4 For antigen dilutions add 5 ml undiluted clarified sap to tube one, adding 5 ml saline or PBS to give a 1/2 dilution. Mix thoroughly. As for serum dilutions transfer 5 ml to the next tube and so on until all eight antigen dilutions are made.

5 Load precipitation tubes with antiserum dilutions beginning with the most dilute (1/2048), 0·5 ml to each of the tubes in row 8 (Fig. 5.3), repeat with dilution 1/1024 to the tubes in row 7 and so on until all tubes in row 1 are filled with 1/8 dilution. Starting with the most dilute sap (1/512) transfer 0·5 ml to each tube in column 8, repeating with decreasing antigen dilutions in each column until all are added. Add 0·5 ml saline in tubes in row 9 and column 9. Mix thoroughly. Where hand pipettes with disposable tips are being used this is best achieved by sucking and expelling the liquids several times.

6 Incubate the tubes, half immersed in the water bath (37°C) to promote mixing by convection (Fig. 5.4).

7 Read the test by holding tubes over the light box, after 1, 2, 4 and 8 min up to 2 hours, rotating tubes rapidly to determine the amount of precipitate and detect trace reactions. A hand lens may help in detecting the latter.

8 Precipitates in tubes are usually ranked as follows:

− = no reaction
± = barely visible precipitation
+ = slight
++ = moderate
+++ = heavy
++++ = very heavy

Two kinds of precipitate may result depending on the shape of the virus particle involved: a granular (somatic) precipitate is formed when spherical viruses are present and a floccular or flagellar precipitate when virus particles are filamentous.

9 The tubes whose contents precipitate most rapidly contain a mixture of antigen and antibody in optimal proportions. The greatest antigen dilution to produce a precipitate is the antigen or virus dilution end-point and the greatest antiserum dilution to give a precipitate is the antiserum dilution end-point or antiserum titre. (See Fig. 5.3.)

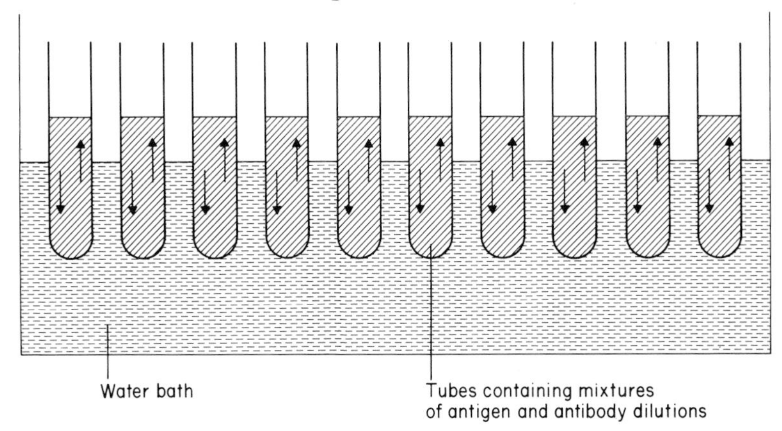

Fig. 5.4. Diagram showing incubation of precipitin tubes to promote mixing by convection.

In addition to the combinations of dilutions of antigen and of antiserum, it is essential to include a small range of the more concentrated dilutions of healthy sap and normal serum (see p. 94).

The antiserum titre is determined by titrating the antiserum against a virus preparation of known concentration of about 10–20 mg l^{-1}. However, if only a sap extract of unknown concentration is available, three titrations are necessary. First, the antiserum titre is determined *approximately* by titrating the serum against one dilution of virus containing sap. Next the virus dilution end-point (see Fig. 5.3) is found using antiserum at four times its *approximate* titre. Finally, the titre is determined accurately using the virus preparation at four times the concentration of its dilution end-point.

The tube precipitin test can also be used to investigate relationships between viruses. Closely related viruses with most of their antigenic determinants in common will behave similarly in precipitation tests with an antiserum to one of them; the titre determined may differ only by one or two dilution steps. If different viruses share only a few determinants the heterologous titres (i.e. the titre of each antiserum to the antiserum of the other), will differ by many steps or there may be no reaction.

Microprecipitin

A disadvantage of the tube precipitin test is its requirement for large quantities of antiserum. The microprecipitin test [13] is a modification of the tube precipitin test which retains its sensitivity and versatility but is much more economical. The principles of the test are the same.

Varying dilutions of antigen preparations and of antiserum are mixed, and the quantity of precipitate formed and the rate at which it is formed are recorded. However, since only single drops of the component antigen and antiserum dilutions are mixed, the quantities used are minimal. A grid titration of the same kind as described for tube precipitation may be set up within a Petri dish (Fig. 5.5). Drops of mixed constituents, accommodated within the spaces of the grid, are then covered with a layer of mineral oil to avoid drying out during incubation.

Materials

1 New clean, unused plastic Petri dishes (in the original technique Formvar coated glass dishes were used. These may be

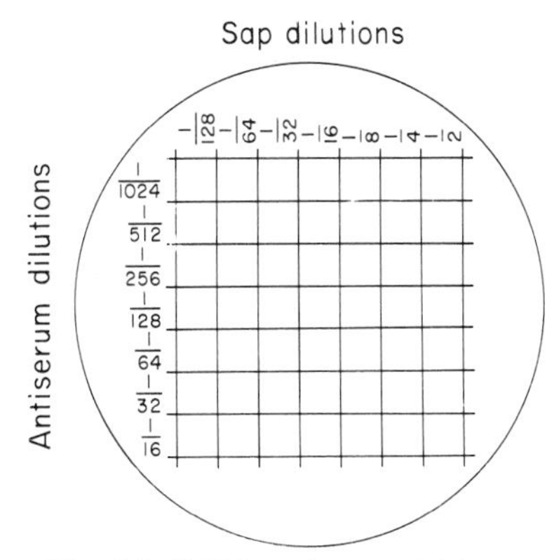

Fig. 5.5. Grid for microprecipitin test.

optically better where observation is to be by dark field microscopy).

2 Wax pencil or graph paper for making a grid within or under the Petri dish.

3 Diluents (0·85% saline or PBS).

4 Suitable tubes for preparation of antigen and antiserum dilutions.

5 Pipettes: adjustable hand pipettes with disposable tips covering a range 5–200 μl, and 200–1000 μl.

6 Antiserum and normal serum.

7 Clarified sap extract containing virus: clarify by centrifugation (5000 **g** for 15 min) or by heating or freezing. The stability of the virus and its concentration in sap should govern the type of clarification procedure adopted.

8 Clarified (as above) healthy sap.

9 Mineral oil (Boots liquid paraffin BP).

Procedure

1 Prepare a grid consisting of 8 mm squares such that the body of the Petri dish base is covered (see Fig. 5.5). This may be done with wax pencil inside the Petri dish, or on separate graph paper to be attached to the base of the dish. The former may be preferable since it is fixed within the plate, and additional locating information may be included. Observations using transmitted light are not prevented, and wax lines also help to keep the drops separate during the test.

Mark the plate in such a way that the orientation of the grid is recorded if a separate paper grid is used.

Prepare a separate record grid in which the location of dilutions of each component can be recorded, e.g. Fig. 5.3. A standard 90 mm Petri dish should, with care, accommodate $7 \times 7 = 49$ separate drops. A full comparison including normal serum and healthy controls will therefore require two plates.

2 Make an appropriate series of dilutions of antigen and antiserum as for tube precipitin, using saline or PBS as the diluent: 10 μl of each dilution of each component (antigen and antiserum) is required for each combination, e.g. to fill all the spaces in Fig. 5.5, a total of 500 μl of each component would be sufficient.

3 Add a 10 μl drop of the appropriate dilution of antigen and antiserum to each space in the grid as designated on the separate test grid (e.g. Fig. 5.3).

4 Complete addition of control drops.

5 Carefully flood the dish with mineral oil from one edge until all is covered, taking care to avoid disturbing the drops. Viscous paraffin oil should be used to avoid spreading of the drops at incubation temperature.

6 Carefully transfer the Petri dish to the incubator and incubate at 37°C for 2 hours.

7 Examine the drops after 2 hours (a) using a stereomicroscope with top light and black background, or (b) using a microscope ($50\times$) with dark ground illumination.

8 After reading reactions after 2 hours, transfer the dish to a refrigerator and read reactions again after storage overnight.

9 Rank the intensity of the reactions as for the tube precipitin test.

As with tube precipitin, the test may be used to determine antiserum titre or antigen dilution end-point. Once antiserum titre is known, the microprecipitin test can provide a simple means of detecting virus. For this purpose the highest dilution of antiserum giving good precipitation should be used. This is both economical and reduces the likelihood of non-specific reactions. Plant sap should be used at the highest concentration which gives good reactions. Tests on one sap extract may be done in one Petri dish using several different antisera to demonstrate presence of several viruses, or the reactions of several virus isolates with one antiserum may be evaluated.

Ouchterlony gel diffusion

A number of tests involve the use of agar or agarose gels through which antigen and antibody diffuse, and precipitation bands form where the constituents meet at suitable concentrations. Such tests separate mixtures of antigens and antibodies by their sizes, diffusion coefficients and concentrations. Thus, they are extremely useful for virus identification from crude sap, clarified sap or from purified preparations. Additionally, the use of antisera which react both to the virus and to proteins present in sap of healthy plants may be tolerated since, in general, the precipitation lines of both will not coincide. The most widely used gel-diffusion test is that devised by Ouchterlony [10] in 1948.

Thin agar or agarose gel layers are prepared on glass slides, or in Petri dishes, and suitably arranged wells are cut to accommodate the reactants. The antigens and antibodies diffuse radially and where they meet in optimal proportions precipitation lines are formed. Because the test is two-dimensional, several virus antigens and antisera can be compared in one test.

Materials

1 Glass slides (*c.* 50 × 50 mm), or flat-bottomed glass Petri dishes, or plastic Petri dishes (new).
2* Sodium azide (avoid contact with skin, never mouth pipette solutions containing this and take care over its disposal).
3 Sodium chloride.
4 Cork borer (No. 3) or commercial gel cutter.
5 Agarose or ionagar No. 2 (Oxoid).
6* Saline (0·85%) or PBS.
7* 0·07 M phosphate buffer, pH 7·0.
8* 0·6 M sodium borate buffer pH 7·5 (plus thioglycollic acid).

Procedure

1 Mix 0·6–1·0 g of agarose or ionagar No. 2, 0·85 g NaCl and 100 ml distilled water or phosphate buffer. Bring to the boil and allow to cool to approximately 50°C then add 0·02 g sodium azide.
2 Maintain agar in water bath to prevent setting. (The properties of agar or agarose change when repeatedly heated, pour plates with preparations heated only once.)

*See Appendix.

3 Pipette (not mouth pipetting) an appropriate amount of agar onto glass slides (*c.* 7 ml) or Petri dishes (*c.* 10–20 ml).
4 Allow agar to solidify, standing slides or dishes on a level surface.

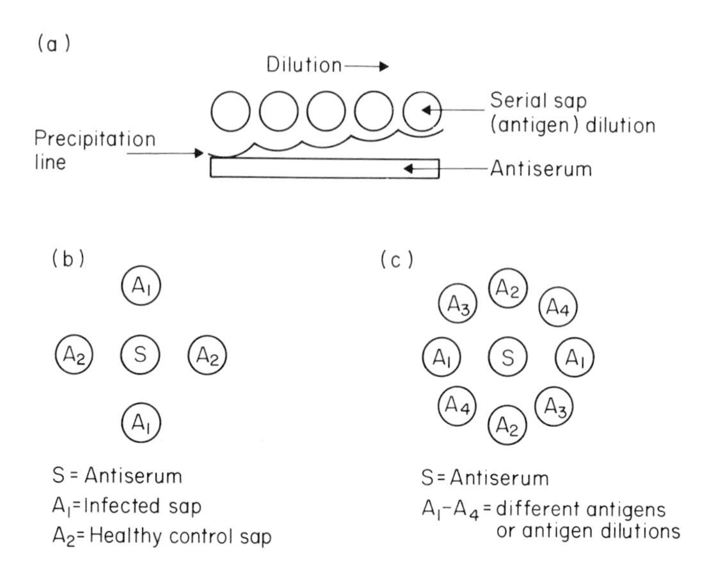

Fig. 5.6. Arrangement of wells for gel diffusion tests. (a) Useful configuration to determine antigen dilutions (which will give clear precipitation lines) before proceeding to comparative tests. (b) A simple confirmatory arrangement which can be repeated several times within a Petri dish using different antiserum and sap dilutions. (c) A more complex array of antigen wells around a single antiserum well for comparison of relationships between antigens.

5 Poured slides or dishes may be stored for a limited period provided they are kept moist and cool. Place slides in a humid chamber, or put damp filter paper under lids of Petri dishes and store at 4°C in a refrigerator.
6 Using a cork borer, cut wells in the gel according to a predetermined pattern (Fig. 5.6). The distance between wells should be 2–4 mm (when a commercial gel cutter is used, several wells, at fixed spacings are cut at once). Remove gel plugs using suction. This is best achieved using a piece of thick-walled rubber or silicone rubber tubing of an appropriate diameter coupled to a vacuum water pump. The outside diameter of the tubing should be a little less than that of the wells, and the end should be cut neatly across with a sharp knife. Care in extracting the agar plug is essential to avoid damaging the edges of the wells.

7 When glass Petri dishes are used it is advisable to seal the bottom of each well using a small quantity of molten agar. This is not necessary if new clean plastic dishes are used, or if glass surfaces are precoated with Formvar (1% solution) or are siliconized with dimethyldichlorosilane solution (see Appendix). Wells should be cut in agar immediately before use to avoid collection of unnecessary liquid in them.
8 Dilute the antiserum with saline or PBS: it may be necessary to prepare several dilutions.
9 Extract the sap in a minimum of buffer (phosphate or borate) (or make dilutions of known concentrations of antigen). Use healthy sap of the same plant as a control.
10 Add reactants to wells in the prescribed pattern, and note the arrangement.
11 Incubate the dish at room temperature ensuring that the agar and well contents do not dry out, since changes in concentration of the reactants may cause artefacts. Keep tests moist by enclosing a damp filter paper circle inside the Petri dish lid, or covering the gel surface with a layer of liquid paraffin or light mineral oil.

Interpretation of results

Precipitation lines often become apparent within 24–48 hours, but incubation may need to be over a longer period (1–2 weeks). Possible reactions which may be observed are as follows.

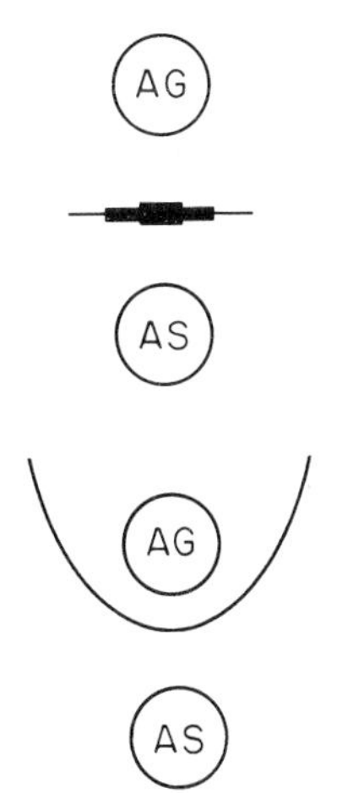

A straight line equidistant from antiserum and antigen will often occur when the antiserum contains antibodies against fraction 1 protein, which is present in both healthy and diseased sap, and diffuses at the same rate as antibody molecules.

Lines nearer to the antigen well, curving around it are characteristic of viruses diffusing more slowly than the antibodies.

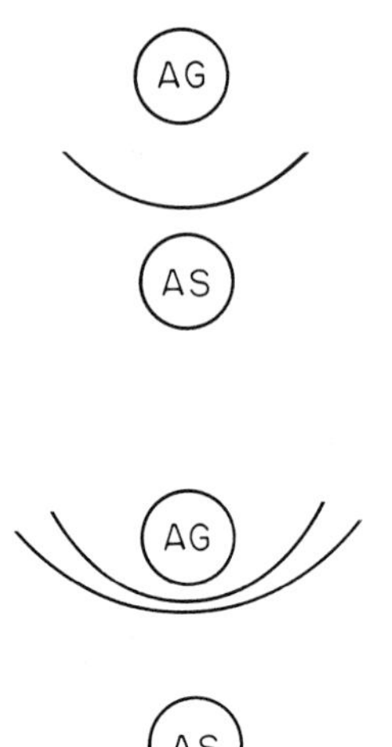

Lines nearer to the antiserum well may indicate that rapid diffusion of virus fragments has occurred.

Double lines may indicate the presence of two components in the antigen well to which there are antibodies in the antiserum, e.g. plant protein and virus. Alternatively a blurring of the lines may suggest that only one antigen/antibody reaction is taking place but that the reactants are not present in optimal proportions and there is antibody excess.

Gel diffusion tests are most frequently used to compare the reactions of several possible different antigens with one antiserum. In this way the relationships between viruses or virus strains can be demonstrated. The reactions in such tests can be complex but the following are common.

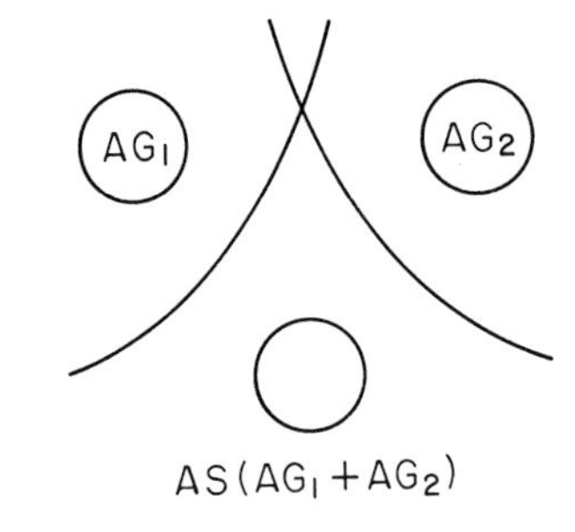

When the antigens have no common determinants, although the antiserum contains antibodies to both, the precipitation lines will cross.

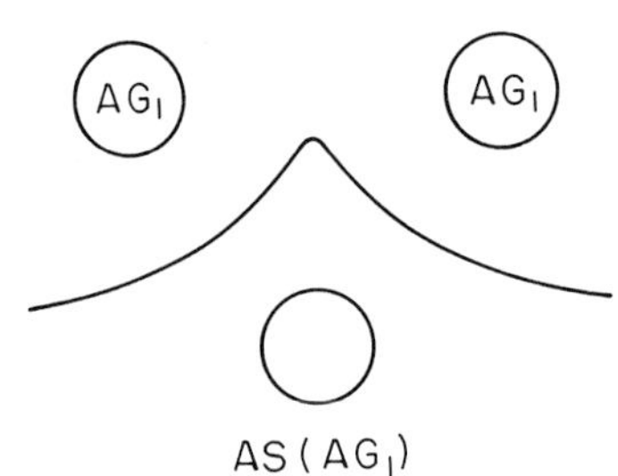

When identical antigens react with their homologous antiserum the ends of the precipitation lines join and do not cross.

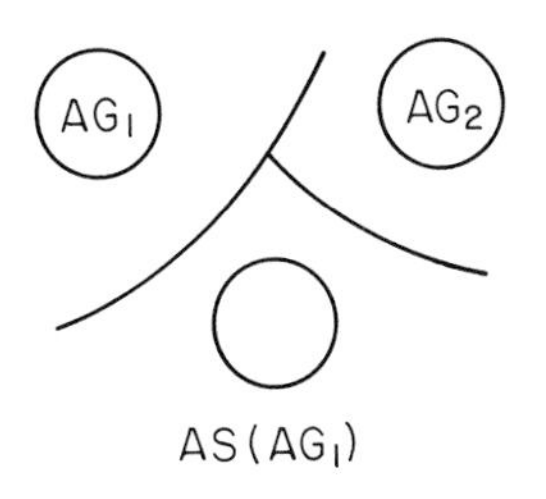

When the antigens have some determinants in common, e.g. with related virus strains, and the antiserum is prepared against one of the antigens a spur will form where the lines of precipitate meet. Antibodies specific to the homologous antigen diffuse through the heterologous antigen precipitin line.

Gel diffusion should be done wherever possible using the reactants at optimal proportions, and at approximately constant temperature. If optimal proportion ratios are not known then a range of dilutions should be prepared. If temperatures fluctuate, multiple precipitation lines may be formed.

One major disadvantage of gel diffusion tests is that they are difficult to use with most viruses which have filamentous particles and do not readily diffuse through gels. Some will diffuse through low concentration (0·5% agar) gels; alternatively, particles may be broken into smaller fragments which will diffuse by ultrasonic vibration or use of detergents. However, fragmentation may change the antigenic character of viruses exposing antigenic determinants which are normally internal.

Agglutination Tests

Tests which involve the formation of precipitates—in tubes, drops or in gels—although sometimes used for routine diagnostic use, may be considered rather complex and more appropriate for quantitative or comparative purposes. Some of the tests involving agglutination have found more widespread use in diagnostic situations although others for various reasons have not become popular. The term 'agglutination' is used instead of precipitation when the size of the reacting antigenic particle is approximately equal to that of a cell. Perhaps the simplest serological test used to detect plant viruses is the chloroplast agglutination test. Mixtures of a drop of antiserum and a drop of sap from a suspected virus-infected plant are made on a glass slide. Control mixtures of antiserum and healthy sap and normal serum and 'infected' sap are also made. When the virus particles and antiserum react together then the chloroplasts from the sap either coprecipitate or agglutinate.

The chloroplast agglutination test can only be used with viruses with elongated particles which occur in high concentration in the plant sap, such as PVX and TMV. Such viruses can usually react over a wide range of antigen/antibody ratios so that preparation of dilution series to achieve optimal proportions is not required. The chloroplast agglutination test is comparatively insensitive and should always contain controls, but it serves as a useful field test in some situations.

A much more sensitive agglutination test is that known as the passive haemagglutination test (passive, since on unsensitized red blood cells there are no receptors for plant virus antigens). Sheep red blood cells are sensitized usually with antibodies (but sometimes conversely with antigen) in a separate process. When such sensitized cells are mixed with extracts containing virus (or antibody), haemagglutination takes place. The test is performed in plastic agglutination plates, a positive result being indicated by the covering of the entire bottom of the well with a thin layer of red blood cells. Densely packed cells in the centre of the well cavity indicate a negative result. The passive haemagglutination test is reported to be much more sensitive than the tube precipitin test. To achieve full sensitivity, however, fresh erythrocytes, less than 18 hours old, must be used.

Another sensitive agglutination test is that involving bentonite flocculation. As with sheep red blood cells, sodium bentonite particles are sensitized with antibody by incubation with diluted antiserum. Flocculation of bentonite particles when mixed with sap extracts indicates the presence of virus. In the absence of virus, bentonite particles remain in suspension for several hours. The bentonite flocculation test is perhaps more prone to non-specific reactions, and strong inhibition of reaction may be caused if antigen is present in excess.

Latex test

Perhaps the most versatile of the more sensitive agglutination tests is the latex test [1]. The test offers a quick sensitive confirmatory test for establishing virus identity (although, as with the passive haemagglutiation and bentonite flocculation tests, it can equally easily be used to detect antibody presence).

In preparatory procedures the specific gamma globulin must be purified, and then used in a separate process to sensitize latex particles. When the homologous antigen is mixed with suspensions of such particles, the latex aggregates (Fig. 5.7). The test has particular value since

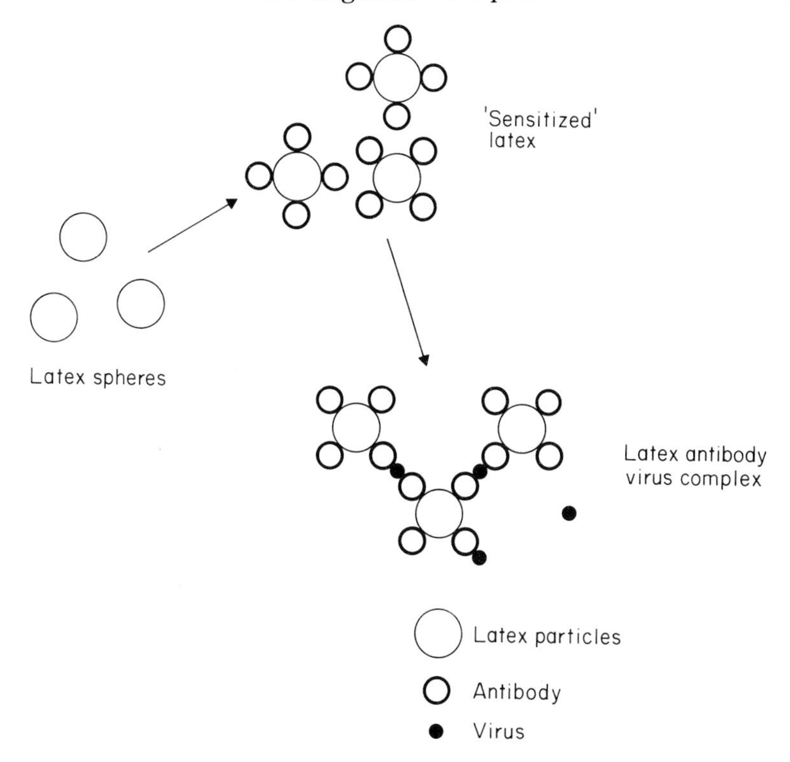

Fig. 5.7. The latex test has several advantages: (i) sensitized latex can be stored or distributed to outstations, and it is (ii) 10–100 times more sensitive than slide agglutination, (iii) independent of virus morphology, (iv) moderately economic of antiserum and (v) quick and simple to carry out.

the preparatory sensitization of latex suspensions may be undertaken at a central laboratory and then distributed to less well-equipped outstations where it can be stored at 4°C until used in testing.

Materials

For preparative and test procedures.

1 Antiserum (virus specific).

2 Latex (Bacto-latex 0·81, Difco Laboratories).

3 Black glass test slides (Orthopharmaceuticals Ltd).

4 Neutral saturated ammonium sulphate solution.

5* Saline (0·85%).

6* Tris HCl buffer, pH 7·2.

*See Appendix.

7 Polyvinyl pyrrolidone (PVP) mol. wt *c.* 44,000.
8 Dialysis tubing.
9 Sodium azide.

Preparation of gamma globulin

1 Dilute 0·5 ml antiserum with 9·5 ml distilled water.
2 Add (whilst stirring) 10 ml neutral saturated ammonium sulphate solution.
3 Incubate at room temperature for half an hour.
4 Centrifuge at low speed (3000–4000 rpm) for approximately 15 min.
5 Discard supernatant liquid carefully then resuspend precipitate in *c.* 2 ml saline.
6 Dialyse against saline (usually three changes of 500 ml) through afternoon, overnight and during morning.
7 Add saline to give a total volume of 5 ml.
8 Add sodium azide to give a concentration of 0·2% and store at 4°C.

Preparation of sensitized latex

1 Add 1 part of latex to 14 parts saline.
2 Add 10 ml of diluted gamma globulin (diluted as predetermined for optimum reaction, see below) to 100 ml of diluted latex suspension.
3 Mix well and incubate at room temperature for 2 hours stirring occasionally.
4 Centrifuge for 25 min at 500 **g** (usually less than 2000 rpm; avoid excessive speeds).
5 Decant the supernatant carefully and discard.
6 Resuspend the precipitate in *c.* 15 ml tris HCl buffer (see Appendix) containing 0·02% PVP.
7 Repeat the centrifugation and resuspension processes twice, and finally resuspend the precipitate in one-quarter the volume of the original mixture (5 ml).
8 Add sodium azide to a concentration of 0·1% and store at 4°C.

There is an optimum ratio of gamma globulin to latex which must be found by experiment before proceeding to preparation of stock sensitized latex. When first using a gamma globulin preparation, prepare small volumes of three or four dilutions which can be used separately to sensitize latex and be evaluated before finally using larger gamma globulin volumes, e.g. dilute gamma globulin to the antiserum titre, to a dilution half as high as the titre and to a dilution twice as high as the titre.

Test procedure

(a) Black glass slide method.

1 Make dilutions of plant sap in the buffer containing 0·02% PVP. It is advisable to make at least a limited range of dilutions when first working with a new host/virus combination; dilutions of sap of 1:10, 1:50 are usually adequate.

2 Place 0·05 ml (50 μl) of antigen dilution in the ringed area on the slide.

3 Add 0·1 ml (100 μl) latex suspension.

4 Mix well.

5 Rock the slide gently by hand or on a rotary shaker (120 oscillations min^{-1}) for 20 min.

6 Reactants become visible after 2–15 min as granular white precipitates which can be better seen by observation with a hand lens.

(b) Petri dish method.

1 A grid pattern should be marked on the inside of the base of the Petri dish (1×1 cm squares) using a wax pencil.

2 Drops of antigen dilution (15 μl) (prepared as in (a) above) are placed in the squares.

3 Drops of latex (15 μl) are added.

4 The Petri dish is shaken on a rotary shaker (120 oscillations min^{-1}) for up to 1 hour after which time any reaction should be visible.

It is essential to include appropriate controls since non-specific precipitates may be formed by some plant sap components.

Reading the reaction

Usually most of the latex particles in controls and non-positive reactions are floating freely in solution after shaking, or spread evenly on the slide when left to settle. In positive reactions the flocculation clumps are evenly distributed around the ringed area and the field has a granular appearance. Sometimes with very strong reactions where there is a high concentration of antigen, a dense aggregation of particles may develop in the centre of the ring. This should not be confused with non-specific clumping, which also forms at the centre of the ring but on observation with a hand lens is seen to be an irregular aggregate.

Antisera may differ in suitability for use in sensitization of latex suspensions. In general, antisera with precipition titres of 1/512 or more are best.

Protein A latex-linked antiserum test

A modification of the latex test, the protein A latex-linked antiserum test (PALLAS) [11], allows the use of antisera with comparatively low titres and may overcome the presence of inhibitory serum or plant sap components. Protein A is a bacterial protein which has a strong affinity for the non-specific (Fc) part of the gamma globulin molecule (see Fig. 5.1). Its use as a component linking the latex particle and the gamma globulin molecule may allow more flexibility in the antigen-antibody-latex particle aggregate but its precise effects are not known.

Materials

For preparation of protein A sensitized latex.

1 Protein A (Pharmacia) 1 mg ml^{-1} stock solution in water containing 0·05% sodium azide.

2 Glycine buffer pH 8·2 containing 1% sodium chloride (see Appendix).

3 Materials as per basic latex test.

Preparation of protein A sensitized latex

1 Protein A solution is diluted 1/200 with glycine buffered saline.

2 Add an equal volume of latex suspension diluted 1/15 with saline.

3 Incubate the mixture for 2–4 hours at room temperature then at 4°C overnight.

4 Centrifuge for 25 min at 500 **g** (usually less than 2000 rpm; avoid excessive speeds).

5 Decant the supernatant and resuspend the precipitate in an equal volume of glycine buffered saline.

6 Repeat the centrifugation and resuspension processes twice, finally resuspending in half the volume of the original mixture.

7 Add sodium azide to 0·02% and store at 4°C.

Preparation of protein A latex-linked antiserum

1 Prepare gamma globulin as for basic latex test and dilute to the experimentally-determined optimum in glycine buffered saline.

2 Mix equal volumes of protein A latex solution and diluted gamma globulin.

3 Mix, and incubate for 2 hours at room temperature.

4 Centrifuge, wash and store as in (**4**)–(**7**) above, except that the final sediment should be resuspended in one-quarter the volume of glycine buffered saline plus 0·02% PVP, as in the original mixture.

The test procedure for protein A latex-linked antiserum is as for the basic latex test.

Labelled Antibody Techniques

Many of the disadvantages encountered by the user of the serological tests involving precipitation or agglutination reactions are overcome by labelled antibody techniques, such as the enzyme-linked immunosorbent assay technique (ELISA). A frequent requirement is the routine detection of plant virus presence in large numbers of samples in surveys. Until the development of ELISA, such routine detection of some viruses was impossible except by the employment of many staff and use of many test plants for sap or vector transmission testing. The ELISA test as described in 1976 [6, 16] has the following advantages.

1 Extreme sensitivity.
2 Applicability to large numbers of samples.
3 Economy in use of high cost antisera.
4 Semi-automatable.
5 Quantitative.
6 Independent of virus morphology.
7 Independent of virus concentration.

The use of the ELISA technique for plant virus detection has been reviewed extensively [5, 15] and a number of modifications and variations have been described.

DAS-ELISA

The direct double antibody sandwich (DAS) ELISA (Fig. 5.8) technique is most frequently employed for the detection of plant viruses and, in

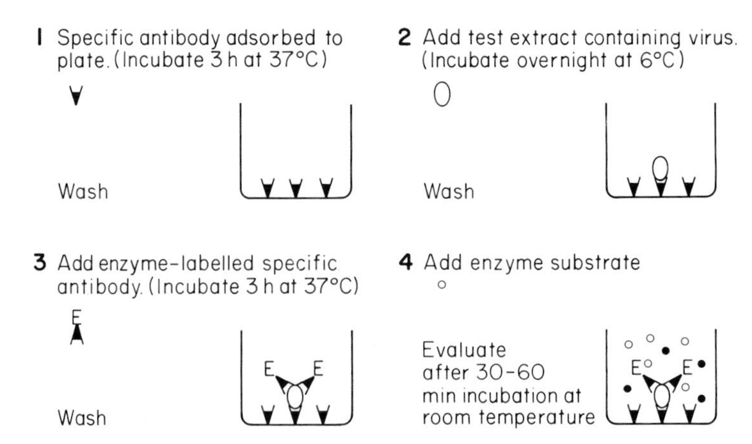

Fig. 5.8. The principles of DAS-ELISA. Colour intensity in the final evaluation is proportional to virus concentration.

most instances, the enzyme used to label the specific antibody has been alkaline phosphatase.

The technique requires (a) purification of gamma globulin and (b) gamma globulin-enzyme conjugation prior to beginning the test process. Prepared globulins and conjugates specific to a small range of viruses can be purchased ready prepared. However, the purification and conjugation procedures are relatively straightforward and once these are carried out within the laboratory most antisera can be used. At its most basic, the ELISA test, including gamma globulin purification and enzyme conjugation can be achieved with simple equipment. However, estimation of gamma globulin concentration during purification requires access to a u.v. spectrophotometer, and accurate evaluation of the end-point reactions of ELISA tests may be limited by visual estimation.

Materials

For preparation of gamma globulin.

1 Antiserum.

2 Neutral saturated ammonium sulphate.

3 Phosphate buffered saline (PBS).

4 Dialysis tubing (7 mm).

5 Pre-equilibrated DE 52 cellulose column: see Appendix for details of equipment and procedure.

6 Dimethyldichlorosilane solution (BDH Chemicals).

7 Suitable tubes for mixing, centrifugation and final storage.
8 Bench centrifuge.
9 Pipettes (1 ml and 10 ml).

Procedure

1 To 1 ml antiserum add 9 ml distilled water.
2 Add 10 ml saturated ammonium sulphate and mix.
3 Leave to precipitate for 30–60 min at room temperature.
4 Centrifuge at low speed (3000–4000 rpm) for 15 min and retain precipitate.
5 Dissolve precipitate in 2 ml of half strength PBS.
6 Dialyse three times against 500 ml half strength PBS (afternoon, overnight and morning dialyses using fresh, half strength PBS each time).
7 Filter through 3–5 ml pre-equilibrated DE 52 cellulose (see Appendix for details of equipment and procedure).
8 Wash gamma globulin through cellulose with half strength PBS.
9 Monitor the optical density of the effluent at 280 nm and collect first protein fraction to elute *or* collect effluent in 1 ml quantities until 12 ml liquid is collected. Check each and retain the ones with the highest optical density at 280 nm (usually the fourth to the sixth).
10 Measure the optical density at 280 nm (use half strength PBS as blank in spectrophotometer) and adjust the strength of the gamma globulin by dilution wih half strength PBS to read approximately 1·4 OD (about 1 mg ml^{-1}).
11 Store in silicone-treated (pre-treat with dimethyldichlorosilane as per manufacturers recommendation) glass tubes at −18°C (but avoid freezing and thawing repeatedly) *or* freeze dry.

Antisera of high titre are preferable but not essential. In most cases, partially purified gamma globulin is used for ELISA although not all workers have gone to the same extremes of purification. In some cases ammonium sulphate precipitated gamma globulins have been sufficiently successful without further purification by cellulose filtration. In this case steps (**7**)–(**9**) may be omitted.

Preparation of enzyme conjugate

The method most widely used for preparation of antibody alkaline phosphatase conjugates is the one-step glutaraldehyde method.

Materials

1 Alkaline phosphatase. This can be purchased as a suspension in ammonium sulphate (Sigma London, Chemical Co. Ltd, Type VII) or as a solution in sodium chloride (BCL). The activity of each such preparation is defined by the supplier in enzyme units per mg. Both these variables are important in preparation of enzyme conjugates: see below.
2 Gamma globulin (1 mg ml^{-1} concentration).
3 Glutaraldehyde solution (25% as prepared for electron microscopy) (BDH Chemicals).
4 Dialysis tubing.
5 Phosphate buffered saline (PBS).
6 Bovine serum albumin powder (Sigma London, Chemical Co. Ltd; Fraction V).

Procedure

(a) For alkaline phosphatase suspensions in ammonium sulphate.
1 Centrifuge 1 ml (equivalent to 5 mg or 5000 units) enzyme precipitate (3000–4000 rpm). Discard supernatant liquid.
2 Dissolve precipitate directly in 2 ml (= 2 mg) purified gamma globulin.
3 Dialyse against three changes of 500 ml PBS.
4 Add fresh glutaraldehyde solution to 0·05% final concentration, mix well.
5 Leave for 4 hours at room temperature during which time a very faint yellow brown colour should develop.
6 Dialyse three times (afternoon, overnight and morning) against 500 ml PBS to remove glutaraldehyde.
7 Add bovine serum albumin to give a concentration of 5 mg ml^{-1} and store at 4°C in the refrigerator.
(b) For alkaline phosphatase solutions in sodium chloride.
Add 0·2 ml (equivalent to 2 mg or 5000 units) enzyme solution to 2 ml gamma globulin (= 2 mg). There is no need for dialysis at this stage in the conjugation process which should follow from stage 4 as above.

DAS-ELISA Test Procedure

Preparatory evaluation of coating gamma globulin and conjugate

Before use in routine ELISA testing the optimal dilution for each gamma globulin and conjugate preparation should be determined experimentally in a test plate. For many plant viruses gamma globulin suspension of 1 mg ml^{-1} (OD 1·4) may be further diluted to 1 or 2 $\mu g\ ml^{-1}$. Concentrations of greater than 10 $\mu g\ ml^{-1}$ are reported to reduce the strength of the virus-specific reaction and increase intensity of non-specific reaction. Conjugate dilutions may be of the order of 1/500 and 1/1000.

The following scheme may be used adopting the routine procedure for DAS-ELISA (Fig. 5.9). Working dilutions of coating gamma globulin and conjugate should be chosen which give the greatest colour at the end of the process, combined with least reaction in control wells. The test plate may also provide an opportunity to determine the best sample extract dilution to use.

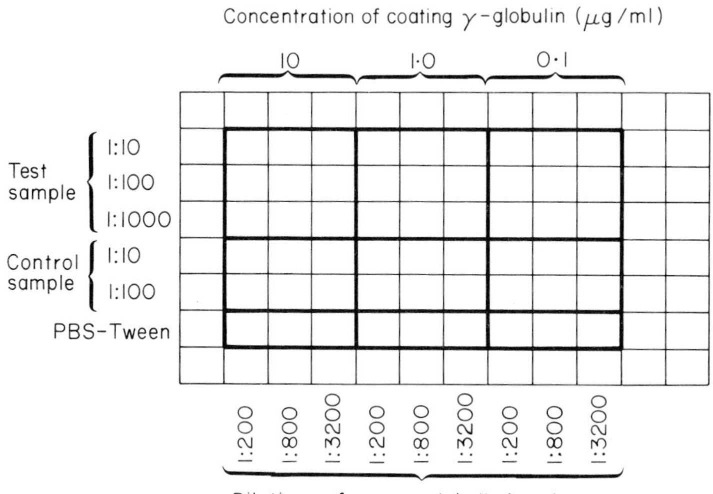

Materials

1 Microtitre plate (polystyrene). A variety of these are available from several manufacturers. The most widely used type, has 8×12 flat-bottomed wells of *c.* 400 μl capacity (see Appendix for list of plate types).

2 Adjustable hand pipettes with disposable tips.
3 Purified gamma globulin.
4 Phosphate buffered saline (PBS).
5* Coating buffer (carbonate buffer pH 9·6).
6* Substrate buffer (diethanolamine buffer pH 9·8).
7 Tween 20 (Polyoxyethylene sorbitan monolaurate). (Sigma London, Chemical Co. Ltd.)
8 Polyvinyl pyrrolidone (PVP) (mol. wt 44,000). (BDH Chemicals.)
9 Ovalbumen (BDH Chemicals).
10 Enzyme-labelled globulin (conjugate).
11 p-nitrophenyl phosphate (Sigma London, Chemical Co. Ltd). (Tablets each of 5 mg chemical are convenient.)
12 Cling film to cover plates.
13 Glassware for dilution of reagents. To avoid loss of specific components this should be scrupulously cleaned and siliconized (using dimethyldichlorosilane solution; see Appendix).
14 Incubator at 37°C.
15 3 M NaOH.

Procedure for DAS-ELISA

1 Prepare a dilution (experimentally predetermined as above) of coating gamma globulin in carbonate buffer.

2 Add 200 μl of diluted coating globulin to each well of the microtitre plate.

3 Cover plate with cling film and incubate for 2–4 hours at 37°C.

4 Discard well contents and wash by flooding wells with PBS containing 0·5 ml Tween 20 per litre (PBS-Tween). Adopt a standard washing procedure throughout the test: flood empty wells with PBS-Tween and stand for 3 min, empty plate and shake dry; repeat three times.

5 Add separate 200 μl aliquots of dilute test sample extract to pairs of wells using a hand pipette and discarding pipette tip between each sample. Samples should be extracted using a pestle and mortar and diluted in PBS-Tween containing 2% PVP plus 0·2% ovalbumin (this should be freshly prepared on the day of use). Each microtitre plate should also contain healthy and known infected control samples and wells filled with sample extraction buffer only.

6 Cover with cling film and incubate overnight at 4°C in a refrigerator, or if speed is essential, at 37°C for 4 hours.

7 Wash plate following procedure outlined in (**4**) above.

8 Add diluted (predetermined experimentally as above) enzyme-labelled gamma globulin, 200 μl per well, filling all the wells in the plate.

9 Cover with cling film and incubate at 37°C for 3–6 hours.

10 Wash plate following prescribed procedure.

11 Add 300 μl aliquots of freshly prepared substrate (p-nitrophenyl phosphate 0·6 mg ml^{-1} *or* 4×5 mg substrate buffer tablets in 30 ml) to all wells.

12 Incubate at room temperature until reactions have progressed sufficiently to allow visualization; usually 30 min is adequate.

13 Reactions may be stopped by adding 50 μl 3M NaOH to each well, whilst shaking the plate to mix.

14 Assess results by *either* visual estimation, *or* measurement of absorbance at 405 nm using a colourimeter.

Comments on component parts of test

Plate coating

Reference to Fig. 5.8 will illustrate the importance of plate coating in the ELISA process. The sensitization of the solid phase polystyrene microtitre plate (or polystyrene tubes, beads or stirring sticks may be used) involves the adsorbtion of gamma globulin proteins in an essentially irreversible hydrophobic interaction. Non-ionic detergents such as Tween 20 prevent this interaction but do not reverse it and are added in later steps to prevent non-specific binding. The microtitre plate provides the most convenient solid phase for ELISA, the most common configuration of 8×12 wells being suitable for insertion into most of the mechanical equipment for washing, plate reading, etc. However, other systems for specific purposes are available. The sensitivity of reaction which can be achieved varies from plates of one type of manufacture to another and in some cases there has been differential and often non-specific variation within plates which must relate to the gamma-globulin coating process. Different manufacturers have developed plates specifically for enzyme immunoassay which may vary in performance and meet specific requirements. Coated plates may, if carefully covered, be stored under deep freeze conditions retaining their activity for many months.

Plate washing

Thorough washing between component stages of the ELISA process is essential to prevent carry over of reactants that are not part of the solid phase double antibody sandwich complex. Usually after the coating, test sample and conjugate incubation processes, plates are washed at least three times with PBS containing 0·05% Tween 20 (PBS Tween) (see Appendix) and often wash liquid is left in the plates for several minutes at each stage. Automatic plate washing machines are available to provide standardization for the process. With some host/virus systems, distilled or tap water has been used for washing without adverse results.

Preparing the test sample

Sample preparation technique should be modified according to the concentration and stability of the virus and the presence of inhibiting host sap components. Most ELISA proponents have prepared the sample in a PBS Tween buffer with the addition of 2% polyvinyl pyrrolidone (PVP) but buffer containing 2-mercaptoethanol has given good results with purified virus preparations in ELISA. Alternatively, diethyldithiocarbamate as an additive also is valuable, especially with some sap suspensions. PVP decreases non-specific 'background' reactions but may also reduce virus-specific reaction. Similarly, the addition of dithiothreitol may decrease both specific and non-specific reactions. Optimum sample extract dilution should be determined experimentally; with high concentration viruses it is possible to use dilute extracts (1/100) to reduce or eliminate non-specific reactions due to host cell components. Long incubation of the sample extract in the plate has been found most effective and incubation overnight at temperatures of 4–6°C has been found most convenient for routine test situations.

Addition of conjugate

Increasing conjugate dilution results in corresponding reductions in the ELISA reaction, but this can be partly compensated for by increasing incubation time. Conjugate incubation is usually at 30–37°C for 3 hours, but regimes of 6°C overnight have also given good results.

Substrate addition

After final washing, substrate is added. This should be freshly prepared at the desired concentration and should be free of colour. Sufficient colour change has usually occurred after 30–60 min when the extent of the reaction can be evaluated. Alkaline phosphatase activity can be stopped by addition of 3 M sodium hydroxide and such 'stopped' plates may be stored (covered with cling film) at 4°C for several hours. Plates may be read visually with a sensitivity of $\pm 0{\cdot}15$ OD_{405} and this may be adequate. However, for more precise evaluation the absorbance of each well component should be determined spectrophotometrically at 405 nm (for the alkaline phosphatase system).

Equipment for ELISA

Provided evaluation of ELISA reactions is visual, and if commercially prepared conjugates are used, the ELISA test can be carried out with the simple equipment listed on p. 116. However, in order to take full advantage of the large sample number capability of the ELISA system, component processes have been mechanized. For loading coating globulin, conjugates and substrates, multichannel pipettes with disposable pipette tips adapted to fill either eight or twelve wells simultaneously can be purchased. A variety of plate-washing machines and plate readers are available, the cost of these relating to their sophistication, and a completely automated system which dispenses and extracts component mixtures, washes and reads plates is available.

The preparation of samples remains the most time-consuming component process in ELISA. The method of preparing extracts must be varied according to the type of sample and must take account of the physical state of the material, the stability and concentration of the virus it may contain and the effects of host plants cell constituents on the ELISA reaction. For succulent, sappy leaves with stable viruses which are not easily mechanically transmissible, a roller press of the kind designed for use with flocculation tests is ideal. Where leaves are less sappy, an automatically-operated pipette can be positioned to deliver buffer into the machine's rollers. Where there is risk of carry-over of mechanically transmissible viruses, small quantities of leaf material may be crushed inside small polythene bags. From raw fibrous material, e.g. graminaceous hosts, freeze or air drying of samples followed by milling has given good results, with the advantage that prepared samples can be subdivided and stored almost indefinitely. For extraction of sap from potato tubers, a dentist's drill provided with a sucking and dispensing diluter, allows extraction, dilution and transfer to the coated ELISA plate in one operation.

Developing ELISA for specific host/virus combinations

Whilst the ELISA procedure is relatively simple, for each new virus/host combination a certain amount of experimentation and practice is required. An appreciation of the limitations of the antiserum available and knowledge of the likely virus concentration and distribution in the plant both in spatial and temporal terms are essential. Experimentation is required to determine optimal concentrations for use of coating gamma globulin, conjugate and sample extract and these may need to be repeated with each newly prepared specific component.

Cross-contamination of wells at the sample incubation and washing stages can cause spurious reactions, although the adoption of standard routine loading and washing procedures helps to interpret such errors. Extreme cleanliness in handling all ELISA components must be exercised, preferably with different equipment being used for conjugate and substrate processes. The re-use of disposable equipment is possible but should be carefully monitored.

Interpretation of results

Quantitative comparison of ELISA values in different microtitre plates is not advisable due to possible plate-to-plate variation in sensitivity. Within each plate therefore, appropriate controls should be included for reference. Where non-specific reactions are low and specific reactions high, plate reading is straightforward. Difficulty in interpretation arises when the range of non-specific and specific reaction values overlap. In such instances it may become necessary to include a large number of known healthy control samples and to determine statistically a threshold level for infection. Several authors have used the mean value for healthy controls plus three times their standard deviation ($\bar{x}+3$ SD) to establish thresholds. Alternatively, values more than twice those of healthy controls have been considered infected.

It is generally agreed that the direct DAS-ELISA test is highly specific and heterologous virus strains may not be detected. The reason for the specificity is not known, but it has been suggested that in the conjugation process there is some impairment of serological activity which is more evident in tests on heterologous than homologous strains. Such specificity can be disadvantageous in scheme or survey situations where strain identity is not important and it is essential to detect all virus strains which occur. The specificity may be overcome by mixing antisera to known strains but there still remains the risk of missing undiscovered strains. Indirect ELISA methods (see below) are not so strain specific, but are more complex and are not described here in detail.

Modifications of the ELISA process

A number of modifications of the double antibody sandwich ELISA technique have been described and used for detection of plant viruses, each proponent claiming some advantage from their technique. Indirect ELISA systems follow a double antibody sandwich process except that

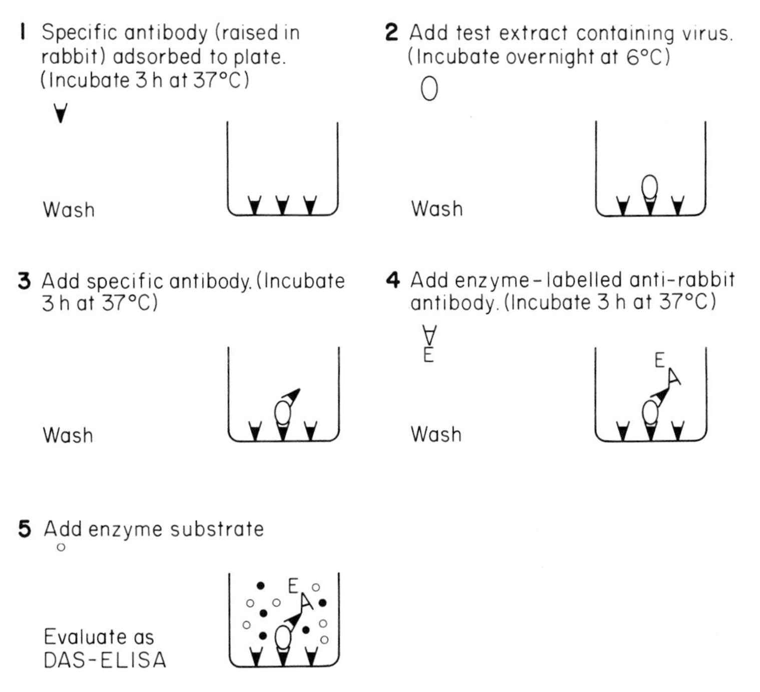

Fig. 5.10. Indirect ELISA sample protocol.

the second layer of specific antibody raised in rabbit is not enzyme labelled, the label being introduced as a conjugated anti-rabbit gamma globulin in a further step (see Fig. 5.10). This, it is claimed, allows the full binding property of the specific gamma globulin to be used, giving greater sensitivity and overcoming the extreme specificity of DAS-ELISA.

Another modification [14] of ELISA uses C1q (a component of complement obtained from bovine serum) to trap virus antibody aggregates (see Fig. 5.11). Plates are first coated with C1q after which a mixture of infected plant sap and virus-specific gamma globulin (raised in rabbit) is incubated overnight. Trapped virus antibody aggregates are detected by the subsequent addition of enzyme-labelled anti-rabbit gamma globulin followed by substrate. The C1q assay not only offers the advantages of indirect techniques but also employs non-specific coating and enzyme-labelled components. However, the technique is adversely affected by concentrated sap of certain plant species, and where this occurs sap must be diluted. C1q assay may not therefore be suitable for routine application where such non-specific sap reactions preclude dilution in grouped samples.

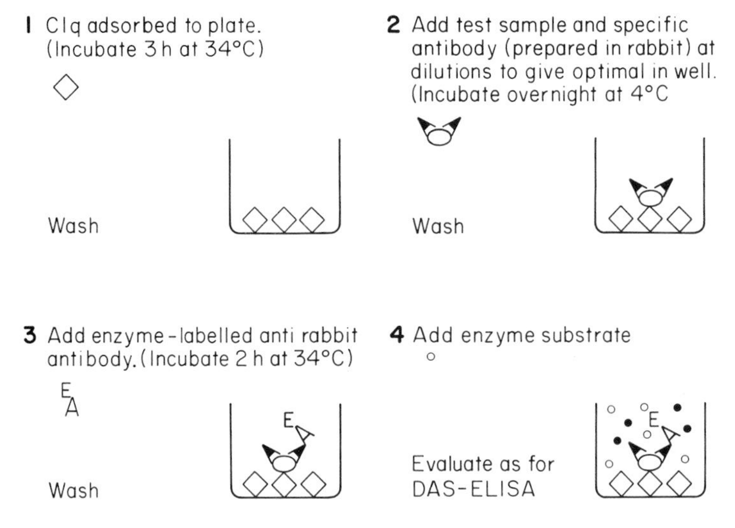

Fig. 5.11. C1q-ELISA protocol.

A further ELISA modification [3] combines the advantages of an indirect assay with those of DAS-ELISA. Antigen is trapped on a solid phase as in DAS but using the $F(ab')_2$ part (see Fig. 5.1) of the gamma globulin molecule (prepared by incubation of gamma globulin with pepsin); trapped virus is detected using unlabelled gamma globulin and this in turn is detected using an immunoglobulin-based enzyme conjugate specific for the Fc portion of the IgG. Pepsin cleavage of the trapping antibody permits the use of a general purpose enzyme conjugate which discriminates between trapping and detective antibody. Disadvantages of the method are that the specificity of the procedure is dependent on the concentration of the second antibody, higher concentration giving decreased specificity. However, lower background reactions are obtained and the procedure may be useful for investigation where the effort or expense of preparing individual virus specific conjugates is not justified.

The use of a fluorogenic substrate (4 methyl umbelliferyl phosphate) has recently been used in comparison with n-nitrophenyl phosphate for the detection of plant viruses. The fluorogenic substrate allowed increases in sensitivity of two to sixteen times with several viruses in leaf extracts, two to four times in tuber extracts for potato virus and allowed more efficient detection of persistent virus in individual aphids and seedborne virus in true potato seed.

References

1. Abu Salih, H.S., Murant, A.F. & Daft, M.J. (1968) The use of antibody sensitised latex particles to detect plant viruses. *Journal of General Virology,* **3,** 299–302.
2. Ball, E.M. (1974) Serological tests for the identification of plant viruses. The American Phytopathological Society, Minnesota.
3. Barbara, D.J. & Clark, M.F. (1982) A simple indirect ELISA using $F(ab')_2$ fragments of immunoglobulin. *Journal of General Virology,* **58,** 315–322.
4. Berks, R., Koenig, R. & Querfurth, G. (1972) Plant virus serology. In *Principles and Techniques in Plant Virology* (eds Kado, C.I. & Agrawal, H.O.). pp. 466–472. Van Nostrand Reinhold, New York.
5. Clark, M.F. (1981) Immunosorbent assays in plant pathology. *Annual Review of Phytopathology,* **19,** 83–106.
6. Clark, M.F. & Adams, A.N. (1977) Characteristics of the microplate method of enzyme-linked immunosorbent assay for the detection of plant viruses. *Journal of General Virology,* **34,** 475–483.
7. Francki, R.I.B. (1972) Purification of viruses. In *Principles and Techniques in Plant Virology* (eds Kado, C.I. & Agrawal, H.O.). pp. 295–325. Van Nostrand Reinhold, New York.
8. Gibbs, A.J. & Harrison, B.D. (1976) *Plant Virology—The Principles.* Edward Arnold, London.
9. Maramorosch, K. & Koprowski, H. (1967) *Methods in Virology,* Vol. II. Academic Press, London.
10. Ouchterlony, O. (1968) *Handbook of Immumodiffusion and Immumoelectrophoresis.* Prog. Allergy. Ann. Arbor. Science Publishers, Michigan.
11. Querfurth, G. & Paul, H.L. (1979) Protein A-coated latex-linked antisera (PALLAS): new reagent for a sensitive test permitting the use of antisera unsuitable for the latex test. *Phytopathologische Zeitschrift,* **94,** 282–285.

12. Regenmortel, M.H. van, (1982) *Serology and Immunochemistry of Plant Viruses.* Academic Press, London.
13. Slogteren, D.H.M. van, (1955) *Serological microreactions with plant viruses under paraffin oil.* Proceedings of the Second Conference on Potato Virus Diseases, Lisse Wageningen, 1954, pp. 51–54.
14. Torrance, L. (1980) Use of bovine C1q to detect plant viruses in an enzyme-linked immunosorbent-type assay. *Journal of General Virology,* **51,** 229–232.
15. Torrance, L. & Jones, R.A.C. (1981) Serological methods in testing for plant viruses. *Plant Pathology,* **30,** 1–24.
16. Voller, A., Bartlett, A., Bidwell, D.E., Clark, M.F. & Adams, A.N. (1976) The detection of viruses by enzyme-linked immunosorbent assay (ELISA). *Journal of General Virology,* **33,** 165–167.

Appendix

Serodiagnosis

Sources of antisera

Antisera are produced at research stations and in research laboratories, usually as a by-product of experimental work. Such antiserum production involves considerable time and resource input and researchers are understandably loth to give preparations away. However, in small quantities some may be made available for serious studies.

Similarly small quantities of many antisera can be obtained by those beginning proper investigations by the American Type Culture Collection, 12301 Parklawn Drive, Rockville, Maryland 20852-1776 USA. (A catalogue of items available may be obtained on request, but the administrators of the collection require evidence (e.g. a letterhead) which shows that requests come from a properly equipped laboratory.)

See also p. 128 for manufacturers of ELISA reagent kits (purified gamma globulin and enzyme conjugated gamma globulin).

Gel diffusion

0·5 M sodium borate buffer pH 7·5

Dissolve 31 g boric acid crystals (H_3BO_3) in 600–700 ml distilled water containing 50 ml M NaOH. This may require heating slightly over a water bath. Make up to 1 litre with distilled water. Add M NaOH dropwise to raise pH to 7·5.

Thioglycollic (0·1%) acid may be added to buffer immediately before grinding leaves.

Staining and preservation of plates

When reactions are clear and complete, wash plate in buffer (PBS) or distilled water, making sure that well contents are thoroughly removed. Stain in amido black solution prepared as follows:

0·1 g amido black 10B;
45 ml acetic acid (12%);
45 ml 1·6% sodium acetate;
10 ml glycerol.

Periodically (1, 5 and 10-min intervals) check depth of staining. Destain to required level using 2% acetic acid.

Latex test

Tris-HCl buffer pH 7·2

1 Prepare 50 ml of a 0.2 M solution of tris (hydroxymethyl) aminomethane (1·21 g in 50 ml distilled water).
2 Add 40 ml of a 0·2 M solution of HCl (8 ml N HCl made up to 40 ml with distilled water).
3 Dilute the mixture to 400 ml with distilled water.

Glycine buffer pH 8·2

1 Prepare *50 ml* of 0·2 M glycine (0·75 g in 50 ml distilled water).
2 Add 2 ml of a 0·2 M solution of sodium hydroxide (0·4 g in 50 ml distilled water).
3 Dilute the mixture to a final volume of 200 ml with distilled water.

DAS-ELISA

Filtration of gamma globulin using DE 52 cellulose columns

Materials
1 10 ml glass syringe (non-luer lock).
2 DE 52 cellulose (Whatman Ltd).
3 Glassfibre paper: Whatman GF/A (Whatman Ltd).
4 Phosphate buffered saline.
5 Gamma globulin preparation.

6 Short tube (to fit syringe nozzle) and clip to close tube.
7 5 ml tubes (in rack) to collect effluent.

Procedure

1 Use the barrel of the 10 ml syringe, fitted with the short length of tube and the clip to regulate liquid flow.
2 Mix approximately five spatulas of DE 52 cellulose in 10 ml of PBS at twice the usual strength (16 g NaCl $litre^{-1}$, etc.).
3 Use the acid fraction of the buffer to bring down the pH of the slurry to 7·4, i.e. 0·5 M KH_2PO_4 added dropwise with stirring.
4 Place a circle of glassfibre paper in the bottom of the syringe. Damp with PBS, close tap. Add cellulose slurry up to the 9 ml level. Open tap and allow cellulose to settle in the column at about 8 ml level. Do not allow surface of cellulose to dry. Place another circle of glassfibre paper on top of the cellulose to keep a level surface when pouring.
5 Run more double strength PBS through the column, checking that the pH of the effluent is 7·4.
6 Run through 20 ml of half-strength PBS and close tap as last liquid reaches surface of cellulose.
7 Carefully add the gamma globulin preparation. Open tap and allow to run through, collecting effluent in 2 ml quantities. When the gamma globulin has run through so that the top disc of glassfibre paper is exposed, add more half-strength PBS until the required number of 1 ml quantities of effluent have been collected (usually ten is sufficient). The whole filtration process takes several (3–4) hours.

Abbreviated method

1 Add 200 μl purified gamma globulin diluted in coating buffer, to each well of the microtitre plate.
2 Incubate for 2–4 hours at 37°C.
3 Wash by flooding wells with PBS-Tween. Leave at least 3 min. Repeat wash two times. Empty plate.
4 Add 200 μl aliquots of the test sample to each or to duplicate wells. Leave at 6°C overnight (or at 37°C for 4–6 hours).
5 Wash plate two times as in (**3**) above.
6 Add 200 μl aliquots of enzyme-labelled gamma globulin to each well. Incubate at 37°C for 3–6 hours.

7 Wash plate two times as in (**2**) above.

8 Add 300 μl aliquots of freshly prepared substrate to each well. Incubate at room temperature for 1 hour, or as long as necessary to observe reaction.

9 Stop reaction by adding 50 μl 3 M NaOH to each well.

10 Assess results by: (a) visual observation; (b) measurement of absorbance at 405 nm.

General use and DAS-ELISA

Phosphate buffered saline pH 7·4

8·0 g NaCl 0·2 g KH_2PO_4 2·9 g $Na_2HPO_4.12H_2O$ 0·2 g KCl	Make up to 1 litre with distilled water to give normal concentration for ELISA use.

PBS may be conveniently handled and stored as a ten times concentrate (adapt above dilution accordingly) which can be diluted as appropriate to 2×, 1× or 1/2× strength as required. Such concentrated stock solutions need no preservative or antibacterial additive such as sodium azide. Diluted PBS however will accumulate bacteria rapidly if kept without preservative. It is best to make each day's supply freshly from the ten times concentrate.

$$\text{PBS-Tween} = \text{PBS} + 0{\cdot}5 \text{ ml Tween 20 litre}^{-1}.$$

Coating glassware with dimethyldichlorosilane solution (BDH)

This chemical gives a water repellent surface to treated glassware. Procedure or use is as follows.

1 Wash glassware thoroughly to ensure freedom from grease (use detergent liberally), rinse and dry.

2 Rinse with dimethyldichlorosilane solution, tip excess away (return to bottle for re-use), and allow to dry.

3 Finally rinse with distilled water to remove traces of HCl formed in the treatment.

Caution: dimethyldichlorosilane gives off an inflammable harmful vapour and should therefore be used in a fume cupboard with appropriate precautions.

Coating buffer pH 9·6

1·59 g Na_2CO_3 2·93 g $NaHCO_3$ 0·2 g NaN_3	Dissolve in 1 litre distilled water.

NaN_3 (sodium azide) is toxic and solutions containing it should not be mouth pipetted. The chemical may also form explosive complexes with metals, so great care should be taken in its disposal and, e.g. use in water bath liquids. The danger from disposal of azide can be minimized by neutralizing it before disposal using ammonium ceric nitrate as follows.

1 To each litre of liquid (containing azide) for disposal, carefully add 100 ml of a 15% aqueous solution of ammonium ceric nitrate.
2 Test for complete decomposition by adding a few drops of ferric chloride solution (13·5 g $FeCl_3 . 6H_2O$ + 2 ml conc. HCl in 100 ml water). If the reaction products remain red after 10 min add more ammonium ceric nitrate.
3 When completely decomposed solutions may then be safely discarded, provided they contain no other hazardous chemicals.

Substrate buffer pH 9·8

97 ml diethanolamine (viscous liquid)
800 ml H_2O (distilled)
0·2 g NaN_3*

Mix together. Then add N HCl drop-wise to bring pH to 9·8. This should bring volume almost to 1 litre. Top up to 1 litre with distilled water.

Manufacturers of ELISA reagent kits

1 Inotech Ag, Zentralstrasse, 34, CH-5620, Wohlen, Switzerland. (Kits of reagents for testing for PVX, PVA, PVM, PVS, PVY and PLRV.)
2 Boehringer Mannheim GmbH, (BCL), Bell Lane, Lewes, East Sussex, BN7 1LG. (Kits of reagents for testing for PVY, PVX and PLRV.)

*See DAS-ELISA coating buffer method for precautionary procedure to be adopted with solutions containing sodium azide.

Manufacturers of microtitre plates

1 Flow Laboratories Ltd, P.O. Box 17, Second Avenue Industrial Estate, Irvine, Ayrshire, Scotland KA12 8NB.
2 Dynatech Laboratories Ltd, Daux Road, Billingshurst, Sussex.
3 Inotech Ag, Zentralstrasse, 34, CH-5610 Wohlen, Switzerland.

6

Electron Microscopy

For many years the electron microscope remained the exclusive tool of the researcher, involving sophisticated and lengthy preparation procedures and an understanding of the technology of the instrument itself. Such complications remain necessary for investigation of many botanical subjects and this, coupled with the high cost of electron microscope equipment has limited their use. Fortunately, many plant viruses are sufficiently stable to survive simple preparation procedures which allow them to be seen in sap extracts using the electron microscope. Additionally, many exist in their host in sufficiently high concentrations to allow them to be detected quickly in crude sap extracts. Thus, with a knowledge of virus morphology, the electron microscope may be used as the first step towards virus diagnosis.

Specimens which are to be observed in the electron microscope must be mounted on a rigid support. This cannot be a glass slide as in microscopy since this would be opaque to electrons. Instead, a fine copper grid or mesh is used, onto which a thin film of plastic or carbon is placed. Such support films are thin enough to be transparent to electrons but strong enough to withstand irradiation. In order to obtain satisfactory images of viruses or other biological specimens in the electron microscope, it is necessary to treat them in some way to enhance their electron-scattering property. The staining method most used in virus detection is negative staining in which electron-dense materials, usually heavy metal salts, surround but do not impregnate the virus particles. Thus, the virus particles stand out as lighter objects against a dark background; hence the expression 'negative' stain.

Whilst negative stains are quick and easy to use they may cause distortion of the particles or may disrupt them. Different viruses react differently to different stains, or to the same stain at a different pH; only experience will show which may be best (see Appendix). For the plant virologist, a limited range of stains which should cover most situations are described in a later section (p. 136).

Access to an electron microscope and its associated facilities is often relatively readily available, and the methods described here allow the casual plant virologist to take advantage of such access. The procedures required to prepare samples for visualization can be executed using only basic equipment. The exception to this may be the preparation of coated electron microscope support grids. In some situations it may be preferable for these to be prepared by the resident electron microscope technician, since the process involves the use of vacuum coating equipment. However, the basic techniques required are not difficult and can be mastered with a little practice. Generalized procedures for support grid preparation are described in the Appendix (see p. 158). Procedures for operation of the electron microscope itself, and for the ancilliary equipment associated with it vary with each instrument and are beyond the scope of this volume.

Visualization of plant viruses in the electron microscope

The electron microscope has most frequently been used in plant virological studies as an aid to diagnosis. The first prerequisite of such use is a reasonable knowledge of virus particle morphology, the recognized groups of plant viruses and their type members (see p. 4). In many host/virus combinations, virus particle frequency—and thus the likelihood of finding particles in an electron microscope preparation—varies with the age of plant, plant part chosen for study, and so on. Thus, some knowledge of host/virus relationships is necessary. Some viruses occur in such low concentration that to find them by simple electron microscope techniques may prove unrealistic.

Having considered the likely range of virus types to be found in the plant material under study, the first practical step to take is to select which plant tissue to sample.

Selecting which tissue to sample

Virus particles are most likely to be found in electron microscope preparations of tissue in which the virus reaches highest concentration. Such tissue should therefore be chosen if possible. There is, however, no rule which defines the disposition of virus within plants; each host/virus combination may differ. All types of plant organs except perhaps meristematic tissue may be invaded by virus, and the investigator should be prepared to utilize any of these. However, the most useful guide to virus presence is the location of symptoms. Leaves showing mosaics and mottles are usually good material to choose. Similarly petals showing 'breaking' of colour usually contain high virus concentrations. With phloem-restricted viruses, such as those of the luteovirus group, the petioles and leaf veins are a good source of virus. For viruses of the tobacco necrosis group, roots may be the only plant part infected.

Having selected the plant part to use, the next step is deciding which method of preparation should be adopted.

Quick methods for sample preparation

There are many different methods for preparation of extracts for virus detection in the electron microscope. A range of basic methods have become accepted, but these are modified according to experience in different laboratories [5]. Some of the methods described below have specific advantages which relate to the kind of tissue used in preparation or to some peculiar property of the virus. All the methods described require certain basic equipment for their execution.

Materials

1 Electron microscope grids provided with suitable support films (see p. 160).

2 Fine watchmaker's forceps: these should either be of the 'crossover' kind which are normally closed or should be provided with rolled elastic bands (or discarded rubber 'O' rings of suitable diameter) so that they can be kept closed (see Appendix and Fig. 6.1).

3 Heavy metal (negative) stains (see p. 136).

4 Clean glass microscope slides.

5 Glass rods with rounded ends.

6 Pasteur pipettes.

In all the techniques described, handling of coated support grids follows a set method. Coated grids are removed from the container or coating mesh using forceps. The grid should be held, coated side uppermost so that the jaws of the forceps hold approximately one-third the diameter of the grid. Rolling the elastic band or 'O' ring forward, the jaws can then be 'locked' and the grid handled with no difficulty (Fig. 6.1). Great care should be taken to avoid distorting the grid during handling, as this leads to breakage of the support film and loss of usable area.

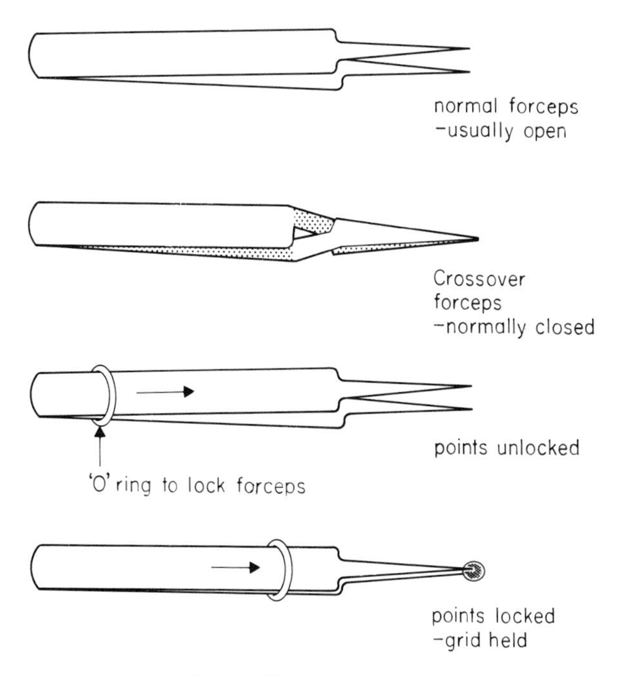

Fig. 6.1. Forceps for handling electron microscope grids.

Leaf squash method

Leaf material is best prepared using this method but other types of tissue, e.g. petals, may also be used.

Procedure

1 Clamp carbon coated grids into pairs of forceps and lay on the bench, carbon coat uppermost, in preparation for receiving extract.

2 Select a small piece of tissue showing symptoms, mosaic or mottle or lesion margin, of about 4–10 mm^2.

3 Squash the tissue on a glass slide in two drops of the chosen stain (see p. 136) (or water, when staining is to be with uranyl acetate, see p. 137) using a rounded glass rod until sap is obviously extracted.

4 Using a fine Pasteur pipette draw up a small amount of the stain/sap mixture and place one drop onto the clamped grid.

5 Carefully remove excess liquid from the grid using a small piece of filter paper touched to the edge of the grid.
6 After 1–2 min drying the grid may be examined in the electron microscope.

The leaf squash method of preparation is the most basic and rapid, but for many high concentration viruses provides excellent results. The disadvantage of the method is that as well as virus, leaf squash preparations contain much cell debris which may prove confusing in the search for virus.

A number of minor modifications of the basic technique have been described. Squeezing of tissue in stain or buffer between two microscope slides gives a more efficient maceration where amounts of tissue are limited. After rotating the macerate between the two slides, they can be prised apart at one end and the sap gathered at the other. This can then be further diluted if necessary so that it is only pale green in colour, then loaded onto the clamped grid as before. In extreme cases the maceration of tissue can be done using the points of forceps actually within a stain drop already placed on the support grid. In certain situations better preparations have been achieved when the support grid is held coated-side downwards and touched onto the prepared stain sap mixture before being drained and dried as usual. This avoids the loading of larger pieces of plant material onto the grid.

Leaf dip method

This technique, like the leaf squash method, is used principally but not exclusively for preparation of leaf samples.

Procedure
1 Clamp support grid in forceps.
2 Load with one drop of stain (or water, if uranyl acetate stain is to be used, see p. 137) using a Pasteur pipette.
3 Cut a small (1 × 3 mm) wedge-shaped piece of leaf and hold in a second pair of forceps.
4 Run the freshly cut edge of the leaf through the drop on the grid over a period of about 30 seconds. Take care not to damage the support film.
5 Remove leaf wedge and drain the grid and allow to dry before examination as before.

As with the leaf squash method of preparation, tissue should be selected which shows symptoms if possible, and where there are distinct lesions the tissue used should include some 'healthy' and some affected. The leaf dip method has the advantage of minimizing the presence of cell debris.

One modification of the technique [3] involves the use of epidermal strips in place of the leaf wedge. These are torn using fine forceps from the underside of the leaf, and allowed to float in the stain drop on the grid. After 30 seconds the epidermal strips are removed and the grid drained as before. Good results may also be obtained by placing drops of stain onto the part of the leaf from which the epidermal strip has been removed, then transferring them to the grid in the usual way.

Other methods

Other tissues containing virus may not be amenable to the leaf squash or leaf dip techniques. For such material a variety of methods for more efficient virus extraction or clarification have been used. Generally, these are modifications of methods used in the initial stages of larger scale virus purification. Grinding small amounts of tissue frozen in liquid nitrogen is an efficient means of extraction of virus from fibrous tissue. Similarly, tougher tissue may be ground in small quantities in a carborundum/buffer paste. Further addition of buffer allows removal of the carborundum by centrifugation, leaving the extracted virus. Some authors have described a means by which petiole exudates are allowed to flow into stain drops on the support grid. Repeated cutting and dipping of the petiole in the stain drop may provide high enough concentration of phloem-restricted viruses. It should be noted however, that immunosorbent electron microscope techniques (see p. 144) allow more reliable detection of such viruses.

Viruses in mushroom sporophores may be prepared for electron microscope examination as easily as those in plant tissue. Sectors of mushroom material cut from the peeled cap (peeled to remove surface debris) may be squashed in a small piece of muslin using a garlic press to extrude sap. This, mixed in twice its volume of negative stain and loaded onto the grid in the normal way, should provide good preparations.

Spray application of virus preparations

Application of prepared and stained extracts by spraying onto the support grid gives droplet patterns which may give a better disposition

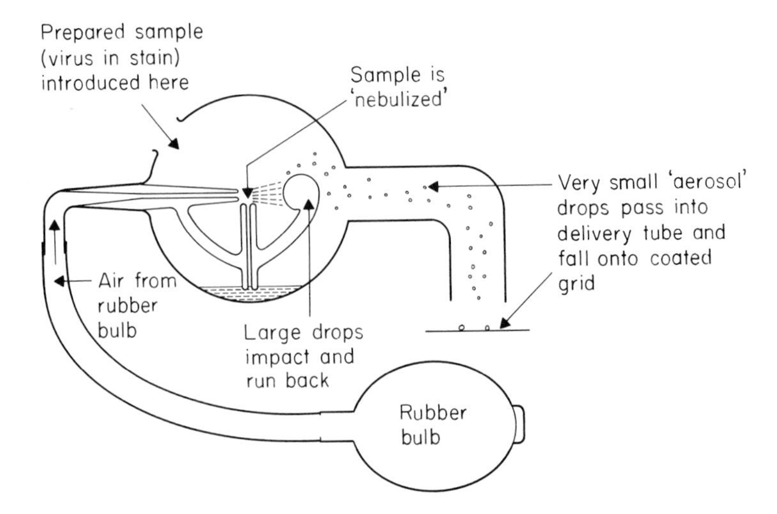

Fig. 6.2. Vaponefrin nebulizer.

of stain around the virus particle than in direct mounted preparations. The Vaponefrin nebulizer (Fig. 6.2) provides an easy means of achieving a spray from small quantities of extract. The horizontal delivery tube should be provided with a downward right angle extension so that the grid can be held horizontally and sprayed from a distance of about 3 cm. Care should be taken to avoid supporting the grid on an extended flat surface or the droplets may be deflected. Similarly, the build-up of an electrostatic charge on the grid may prevent droplet deposition. An alternative to the Vaponefrin nebulizer is an artist's airbrush. These may be adjusted to give droplets of the appropriate size but spraying should be from a greater distance. Care should be taken in the spraying of stained sap extracts, as aerosols of heavy metal stain should not be inhaled. Application by spraying may damage elongated virus particles.

Types of negative stain

Phosphotungstic acid (PTA)

This stain is prepared by dissolving the solid dodecatungstophosphoric acid in glass-distilled water to give a 1 or 2% w/v solution. It may be used at different pHs between 5 and 8, the most usual being pH 6·8. Dropwise addition of N sodium hydroxide solution is used to raise the pH. Prepared PTA is stable in light at room temperature, it can be made up in buffer solutions without being affected by them and is generally regarded as a good 'all round' stain. However, PTA may destroy some kinds of virus particles, or may empty a proportion of initially 'full' ones. It may give poor resolution and variable virus penetration at different pH values.

Ammonium molybdate

Prepare this stain by dissolving the finely-ground solid in distilled water to give a 2% w/v solution. The solution can be used without pH adjustment, but can be amended by adding N sodium hydroxide dropwise to pH values between 3 and 9. Ammonium molybdate is stable in light at room temperature and like PTA can be made up in buffer solutions (except borate) if desired without ill effect. It gives good resolution of virus particles and may be used as a good alternative to PTA. However, ammonium molybdate may give preparations with poor contrast and at low pH values may be difficult to spread on carbon-coated grids.

Uranyl acetate

Normally used as a 1–2% solution of the solid in distilled water without pH amendment, the pH is from 3 to 4 since the solid precipitates at higher pH values. The solution must be kept in a dark bottle, stored in the dark and refrigerated since it precipitates in strong light, and is unstable if heated. In contrast to PTA and ammonium molybdate, uranyl acetate should not be mixed with the tissue during sample preparation (see p. 133), but should be added as a drop wash to the grid containing the virus preparation (prepared in distilled water). Uranyl acetate gives preparations with high resolution and high contrast and provides reproducible results. However, it cannot be mixed with buffers, and may cause clumping or end-to-end aggregation, loss of flexibility of some viruses, and shrinkage of others. It may not be so convenient for use with sap extracts, which may have to be washed after staining to avoid dirty preparations.

Uranyl formate

This stain is prepared by dissolving the solid in distilled water to give a

1–2% w/v solution. The solid is slightly unstable in the light and the prepared solution even more so. Both must be stored cool in the dark, and solutions should not be stored for more than a week if possible. As with uranyl acetate the stain is acid (pH *c.* 3) and should be added dropwise to virus preparations (made in water) already on the grid, and not mixed with virus beforehand. Uranyl formate gives high resolution, high contrast and reproducible results. However, it is unstable and cannot be mixed with buffers. It may cause clumping, end-to-end aggregation, changes in flexibility and shrinkage of different viruses. Uranyl formate is less suitable for sap preparations.

Methylamine tungstate

This stain is available ready prepared from Emscope Laboratories Limited, Kingsnorth Industrial Estate, Wotton Road, Ashford, Kent. The only organic heavy metal salt stain, it is supplied as a 2% solution. It is stable in light at room temperature although, as with all stains it should be stored properly corked in a refrigerator. The stain is used at pH values between 5 and 8, N NaOH being added dropwise for adjustment of pH. Methylamine tungstate can be used in sap extracts in the same way as PTA and need not be added after grid loading. It is a useful general purpose stain providing clean preparations. It is less disruptive than PTA and may be safely used with most viruses.

There are no reviews of the best ways to prepare different plant viruses for examination under the electron microscope. Information concerning experiences with individual host/virus combinations may sometimes be found in the literature concerning specific virus studies. However, such information is limited in extent. For each virus and each host it may be necessary to experiment with different preparation techniques and different stains, indeed variable results may occur in different laboratories using the same technique. Virus in crude sap preparations behaves differently in the electron microscope to purified preparations. The latter may be difficult to spread on the grid with consequent poor results. Pretreatment of coated grids with a drop of 0·002% bovine serum albumin, or 0·02% bacitracin (mol. wt 1411 Sigma London, Chemical Co. Ltd) will overcome this problem. Generally, the behaviour of specific host/virus combinations in EM preparation may relate to their behaviour in other tests or *in vitro* situations. For example, if the plant sap (e.g. that of Rosaceous hosts, grapevine or poplar) contains oxidizing enzymes, or if the virus is particularly labile, then 2% polyvinylpyrrolidone or 0·02 *M* sodium sulphite may be included in the extraction process. Similarly, a greater yield of cucumber mosaic virus particles may be achieved if leaf material is homogenized in 0·05 *M* phosphate buffer, pH 7.

Observation and interpretation of prepared grids

For many viruses with rod-shaped, filamentous or bacilliform particles (Figs 6.3–6.5) recognition in the electron microscope provides no great difficulty. Usually the regularity of particle thickness and length within

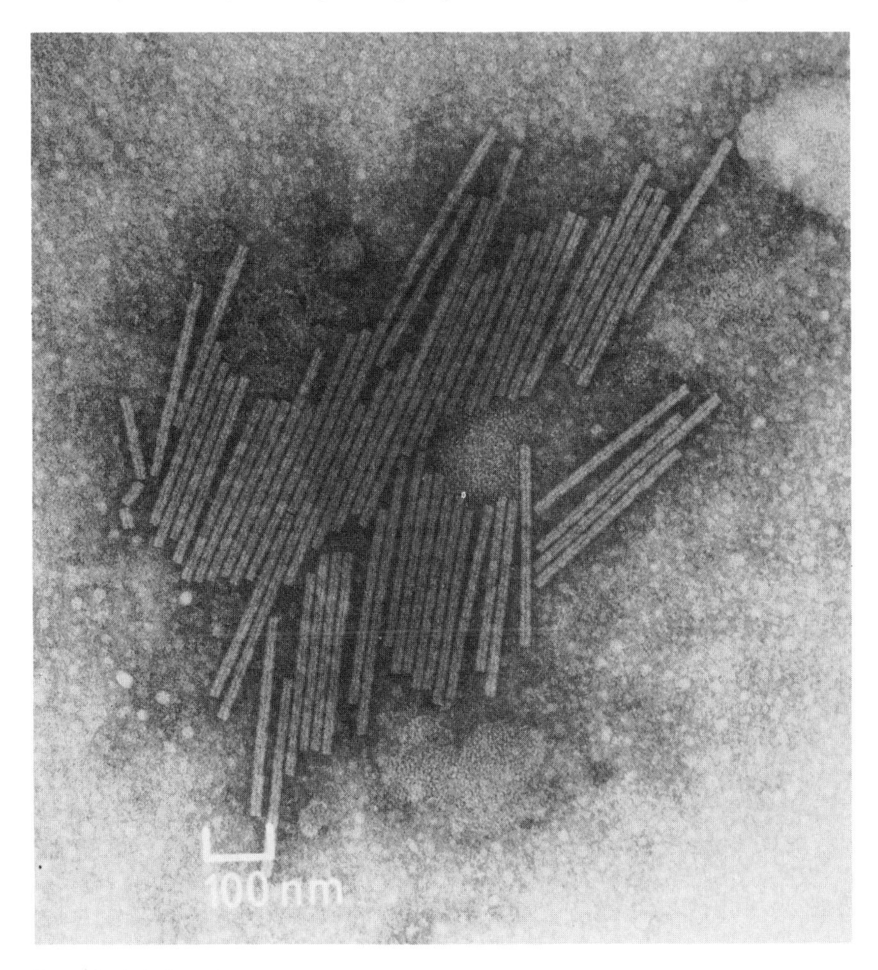

Fig. 6.3. Rod-shaped particles of tobacco mosaic virus (300×18 nm).

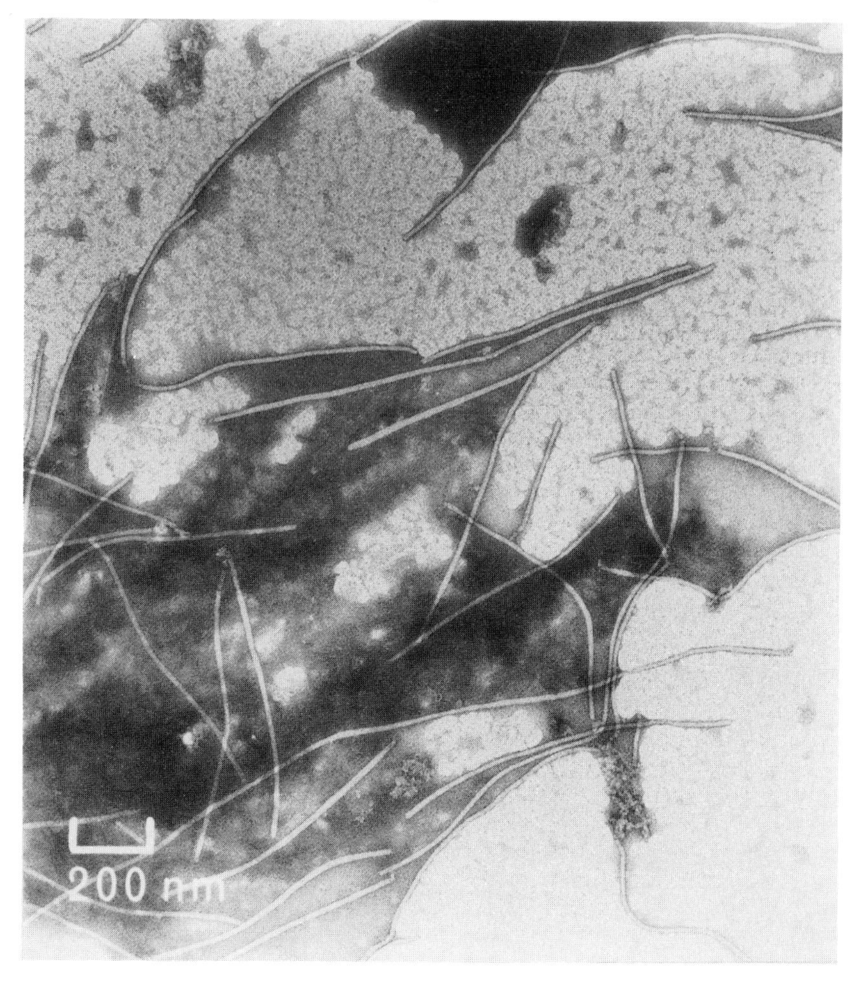

Fig. 6.4. Filamentous particles of potato virus Y (750×11 nm).

a group of particles when seen in the field of view is adequate confirmation of their viral nature. However, even when particles have such distinct morphology there may be confusing artefacts which are particularly common in quick preparations. For example, particle fragments may appear spherical when seen end-on (Fig. 6.6) and others may have different dimensions. In other preparations pieces of the flagellae of some bacteria can be mistaken for filamentous virus particles (Fig. 6.7). However, it is usually the definition of spherical virus particles which presents greatest difficulty, especially when these occur in low

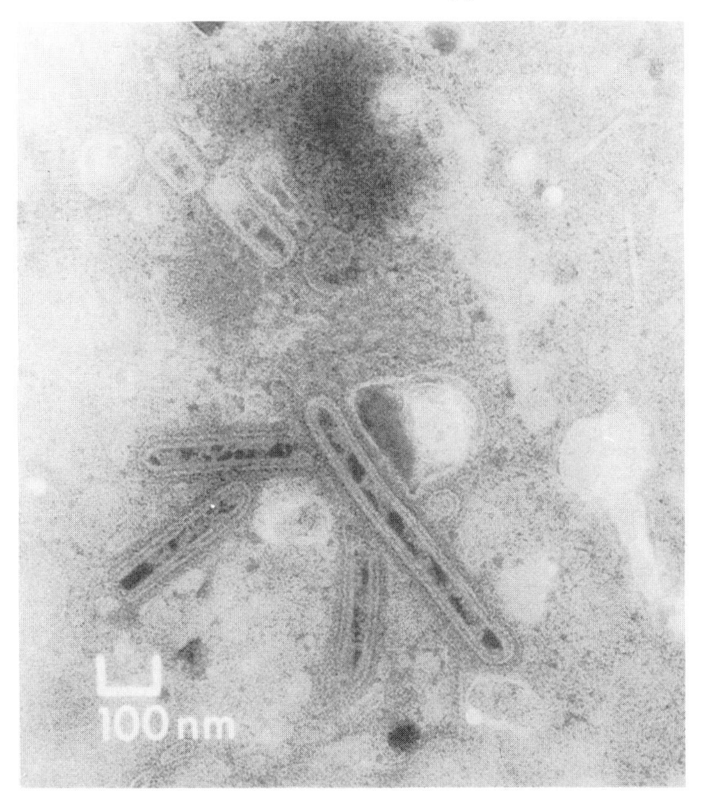

Fig. 6.5. Bacilliform particles of broccoli necrotic yellows virus (380×95 nm). Note broken 'bullet' shapes and part particles.

concentration (Fig. 6.8). A frequent feature of some spherical virus preparations is the mixture of 'empty' and 'full' particles (Fig. 6.9). Parts of plant cell organelles and other debris present in crude extracts may prove confusing. An important criterion is regularity of size and shape, although some filamentous particles tend to break and aggregate end to end.

Measurement of the dimensions of virus particles in the electron microscope may present some difficulties since, even on the most modern electron microscopes, the indicated magnification may vary slightly from one occasion to another. Additionally, virus particles may extend or contract according to the ionic content of the crude extract and with different stains. For reliable measurement, an internal standard, such as tobacco mosaic virus particles, should be included and a frequency distribution of the virus particle dimensions should be prepared.

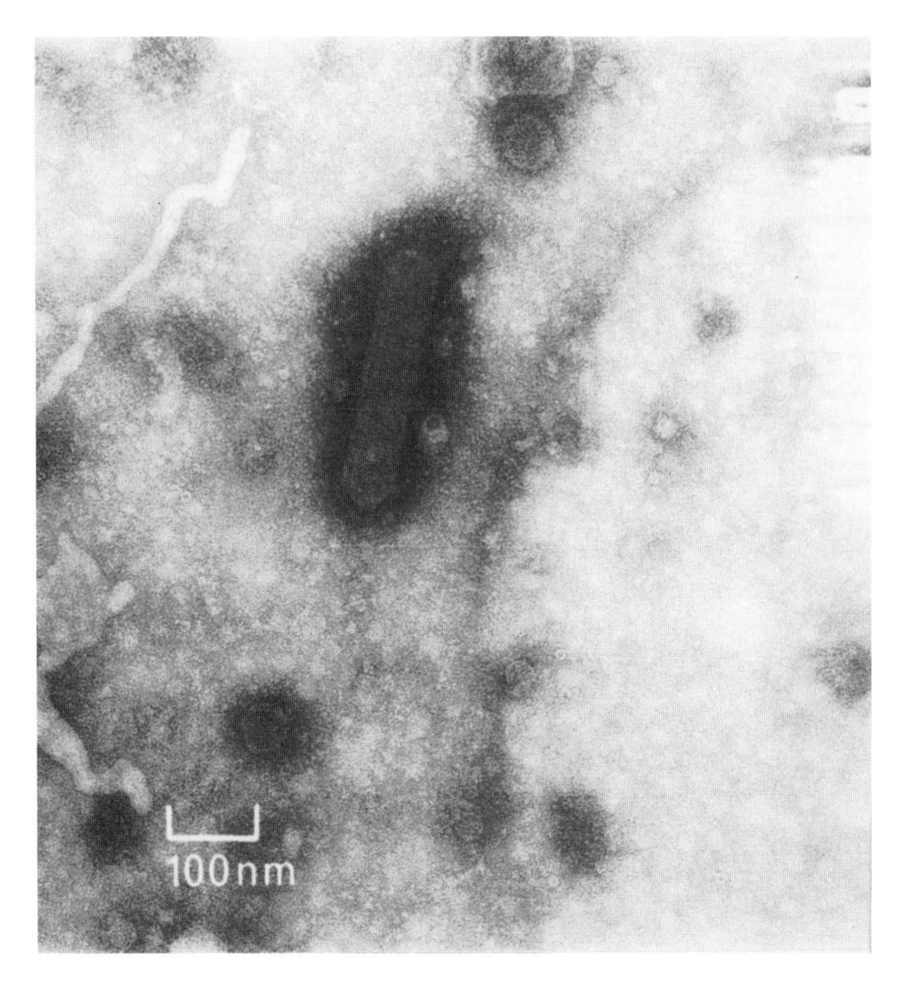

Fig. 6.6. Bacilliform particle and end-on views of particles which could be confused with spherical particles.

In good preparations and with a well maintained microscope it is often possible to see the surface structure of the viral capsid. Often the central core of rod-shaped viruses is penetrated by stain and can be seen clearly. Some groups of viruses characteristically have some of their particles without nucleic acid. Such 'empty' particles are quite distinct.

Quantification of virus in electron microscope preparations requires special techniques. The particle frequency on the prepared grid may bear little relation to that in the sap of the host plant, since much may be

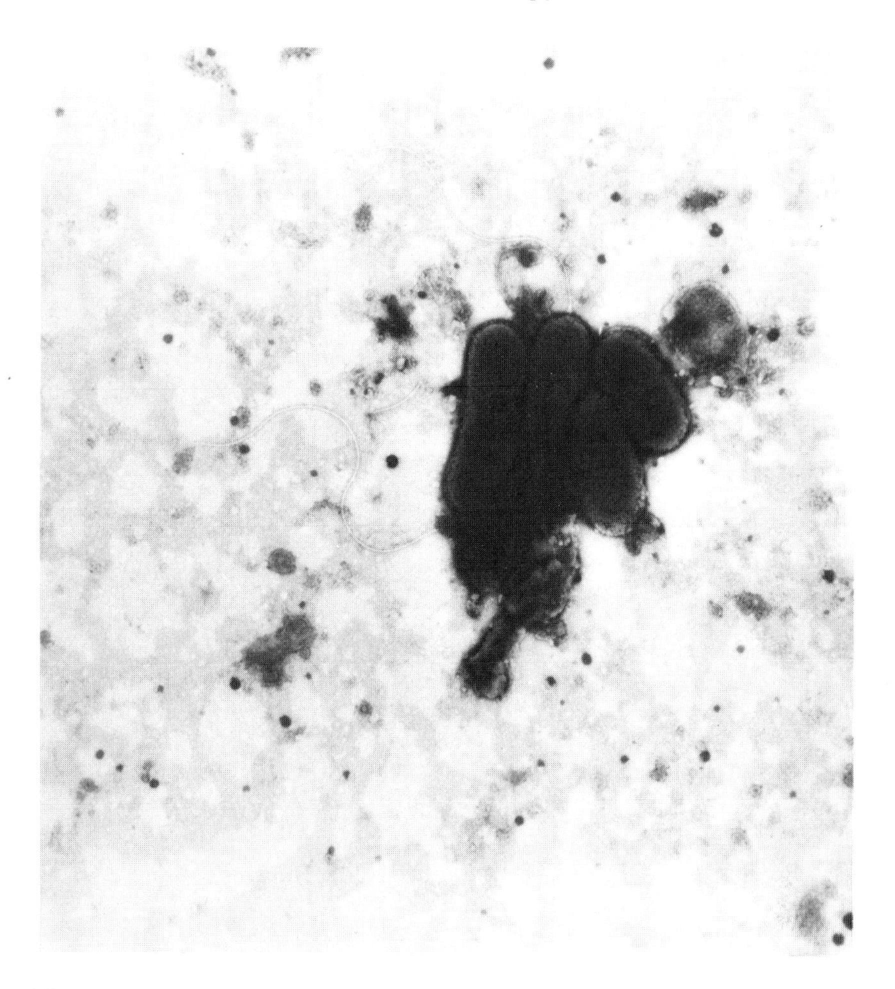

Fig. 6.7. Photomicrograph of uniflagellate bacteria illustrating detached portions of flagellae which may be mistaken for filamentous virus particles.

lost in the process of grid preparation. Methods for quantification of virus particles involve spray application of virus preparations containing known concentrations of latex spheres. The relative frequency of virus particles and latex spheres can then be counted in discrete droplets of liquid on the grid and appropriate calculations made [4]. A simpler method has also been described, whereby the area of the electron microscope's field of view in the microscope binoculars is determined relative to a standard area [6].

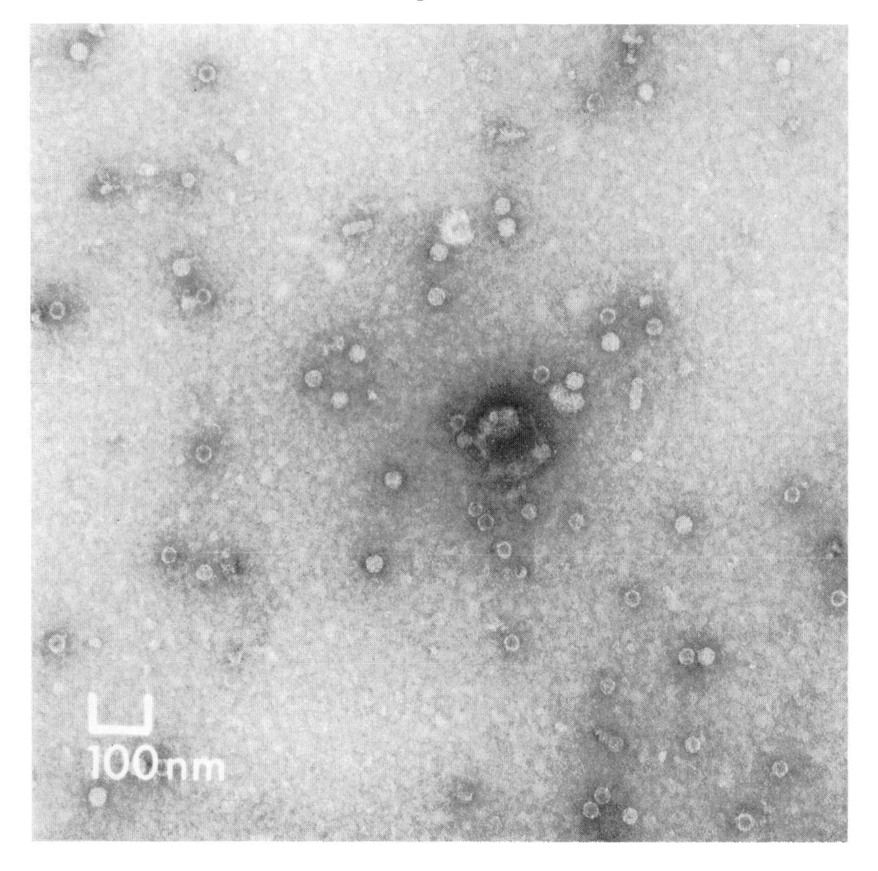

Fig. 6.8. Two spherical viruses of different diameter in the same preparation. Mushroom Virus 1 (25 nm diameter) and Mushroom Virus 4 (35 nm diameter).

Immuno Electron Microscopy (IEM)

The quick preparation techniques described above—when used in combination with appropriate stains and staining procedures—are ideal where virus concentrations within the plant are high and the virus is relatively stable. However, by comparison with other techniques used in plant virus diagnosis, e.g. transmission tests and serodiagnosis, they may be insensitive and since many plant viruses share the same morphology they are non-specific. Immune electron microscope techniques combine the techniques of electron microscopy and serology, involving the detection of complexes of antigen (virus) and specific antibody. Such

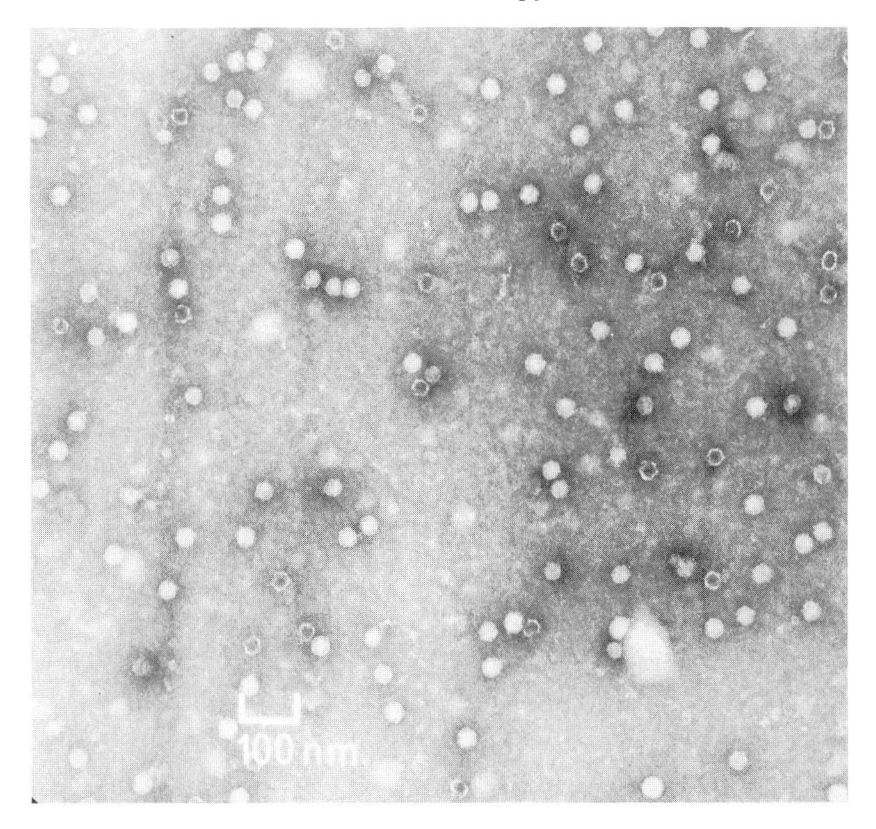

Fig. 6.9. Spherical particles of broad bean stain virus (25 nm diameter). Note 'empty' and 'full' particles.

techniques include the detection of immune reactions in fixed or sectioned material. These will not be described here. Of particular use in the diagnosis of plant virus diseases are those techniques which detect immune reactions in particulate material. Observation of complexes formed between viruses and their specific antibodies in the electron microscope is not a new technique, the pioneer work in this field dating back to 1941. However, techniques of particular value have been described within recent years and new developments are commonplace. Three basic techniques are widely used with only limited variations from one laboratory to another. These are: immunosorbent electron microscopy (otherwise known as 'serum activated grids', 'trapping' or 'the Derrick method' after its author), decoration, and clumping (or 'antiserum virus mixtures'). The development of these methods, and variations from the basic techniques have been extensively reviewed [5].

Immunosorbent electron microscopy (ISEM) (Fig. 6.10)

In this procedure, support films are precoated with specific antibody to which homologous virus particles in the tissue extract become attached and concentrated. Such successive fixing of antibody and antigen allows interpolation of washing stages to remove salts or other undesirable components.

This method is usually used to detect viruses present in host plant sap in concentrations too low to permit detection in conventional EM preparations (see Fig. 6.11). If carefully followed, the technique gives an effective concentration of virus on the grid of from ten- to a hundredfold. The technique can be used with mixed antisera (to 'trap' more than one virus at once) and can be used to compare serological relationships between virus strains, since the efficiency of trapping is related to the degree of homology between antiserum and virus.

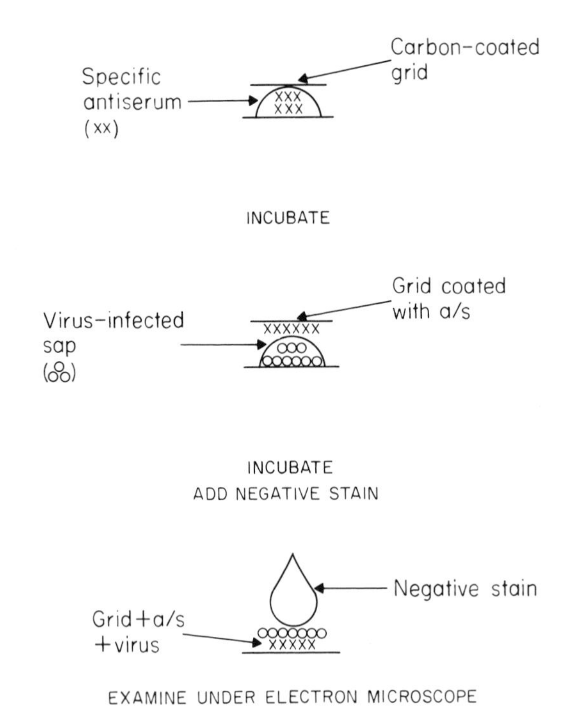

Fig. 6.10. Immunosorbent electron microscopy procedure.

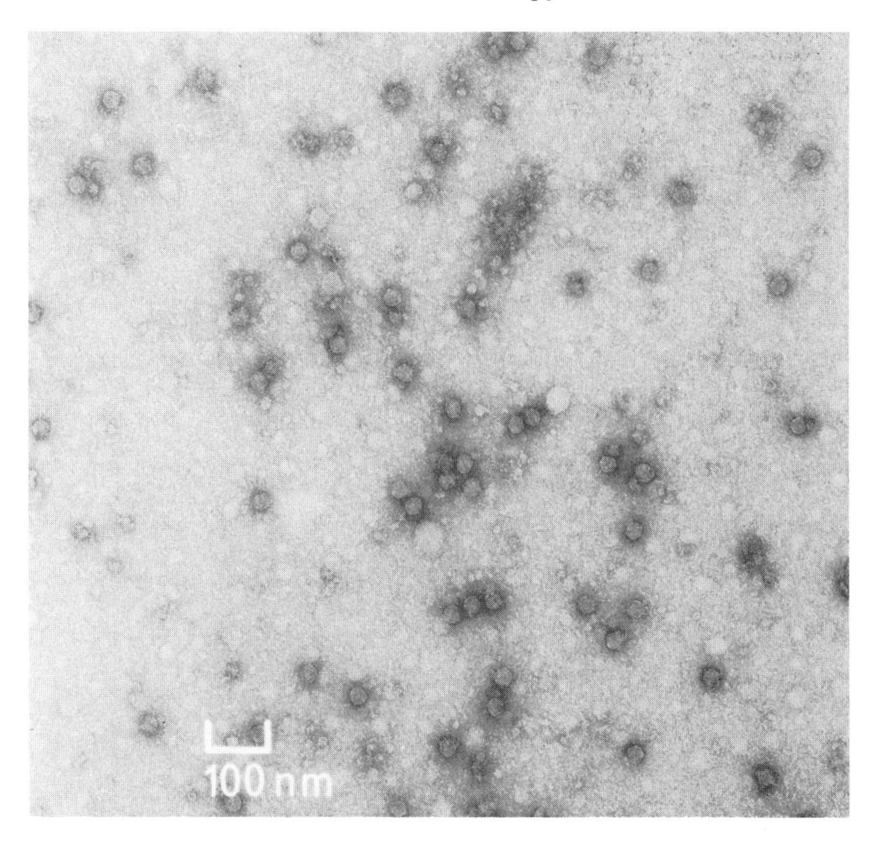

Fig. 6.11. Particles of potato leaf roll virus (24 nm diameter) attracted to the grid using ISEM.

Materials

1 Clean glass Petri dish.

2 'Parafilm' or paraffin wax ('Parafilm' from the American Can Co.).

3 Filter paper.

4 Support grids with carbon film (freshly prepared).

5 0·06 M phosphate buffer, pH 6·5–7·0 (Sørensens).

6 Adjustable pipettes (5–100 μl).

7 Negative stain, e.g. 2% phosphotungstic acid pH 6·8, methylamine tungstate, or 2% ammonium molybdate.

8 Antiserum. Wherever possible, use antiserum which does *not* contain glycerol as a preservative as this reduces staining efficiency and may give poorer reactions. Freshly prepared phosphate buffer (Sørensens 0·06 M pH 6·5) is suitable for dilution of antiserum, although results may vary with buffers of different pH. The degree to which the antiserum should be diluted must be determined by experience. As a general rule, for rod-shaped and filamentous viruses use antiserum diluted to 1/10 the tube precipition titre and for spherical viruses dilution of 1/20 tube precipition titre is often best.

9 Virus preparation. It is important to extract any virus from the plant tissue effectively, then to eliminate as much plant debris as possible. The ISEM technique works satisfactorily with crude extracts prepared by simply grinding small pieces (2×2 mm) of tissue in a drop of buffer, but more sophisticated extraction and clarification may be advantageous. Grinding tissue in buffer containing carborundum, with a suitably shaped piece of glass rod in an Eppendorf tube, followed by low speed centrifugation to remove carborundum and debris, gives good preparations with more difficult viruses.

Procedure

1 Line the inside of the base of the Petri dish with Parafilm, or coat with a thin layer of paraffin wax.

2 Mark base and lid to establish relative orientation.

3 Draw a grid of 1 cm squares on the outside of the lid to locate drops if different preparations are to be examined.

4 Pipette separate 20 μl drops of diluted antiserum onto base of Petri dish. Take care to handle the dish carefully to avoid spreading the drops. Note their position relative to the grid layout (mark on lid).

5 Carefully place a support grid carbon film down on each drop so that it floats on the drop.

6 Cover Petri dish with lid to prevent evaporation and incubate at 37°C. Incubation for 30 min or less is often adequate but some practitioners recommend incubation periods of 3 hours (see below for discussion of incubation times).

7 Wash grids either by floating them on the surface of several mls of buffer (Fig. 6.12), or by flooding dropwise with buffer from a Pasteur pipette, holding grids in forceps. Washing should be thorough to remove excess antiserum.

8 Float each antiserum-coated grid, coated side downwards on

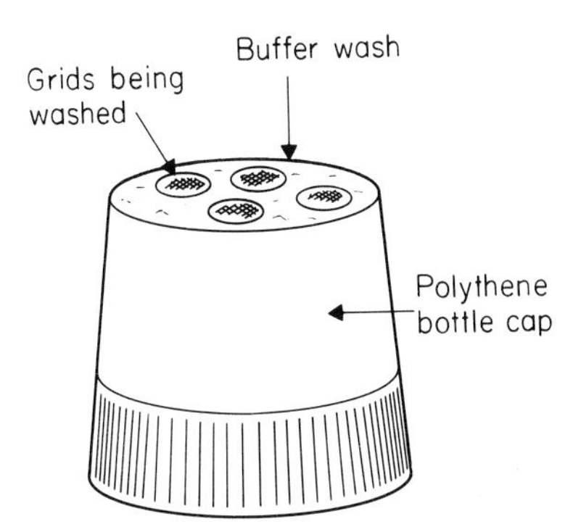

Fig. 6.12. Washing grids for ISEM: agitate periodically by sucking and releasing wash buffer using a Pasteur pipette.

a 20 μl drop of virus extract on Parafilm, or on the waxed surface of a Petri dish (as for antiserum coating in (**5**) above).

9 Incubate grids plus virus in the covered Petri dish for 3–36 hours at 4°C or at room temperature for 1–5 hours (see below for discussion of incubation times).

10 Immediately after draining and blotting the grid dry, stain by placing face downwards on one drop of stain. Drain and blot dry after 30 seconds.

To some extent the objective in using ISEM defines the duration and temperature of the two separate incubation stages in the technique. Whilst there seems little to be gained by increasing the antibody adsorption incubation time, considerable increase in sensitivity may result from longer virus adsorption. Thus, where time is short, in experimental situations, the duration of the incubation of antibody-coated grids on virus extracts may be kept to a minimum. For routine tests, however, increasing the duration of antigen incubation from 3 to 36 hours may give a five- to tenfold increase in the amount of virus trapped. Incubation at lower temperature is said to give greater uniformity of trapping and avoids evaporation of the extract. Incubation at room temperature gives more rapid reactions. Some viruses (rhabdoviruses, ilarviruses and tomato spotted wilt) may deteriorate after prolonged incubation at low temperatures.

Increased sensitivity of the ISEM procedure has been claimed using protein A coated grids [8]. This bacterial protein has a strong affinity for the Fc portion of the gamma globulin molecule (Fig. 5.1). Thus, support films precoated with protein A (0·02 mg ml^{-1} in 0·06 M phosphate buffer pH 7), and subsequently antibody-coated should have all the gamma globulin molecules orientated with their receptive arms outwards and, in theory, the efficiency of virus trapping should be greater. Preliminary results suggest that the increase in sensitivity achieved is modest, but further investigation may yet be required.

Decoration (Fig. 6.13)

As the name of the technique suggests, this procedure involves the decoration, or coating of virus particles already on the grid with a layer

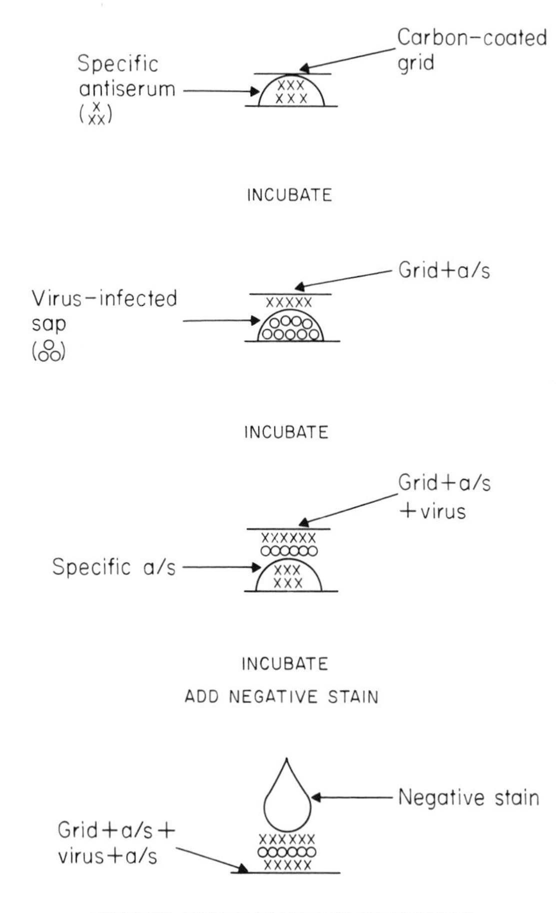

Fig. 6.13. 'Decoration' procedure.

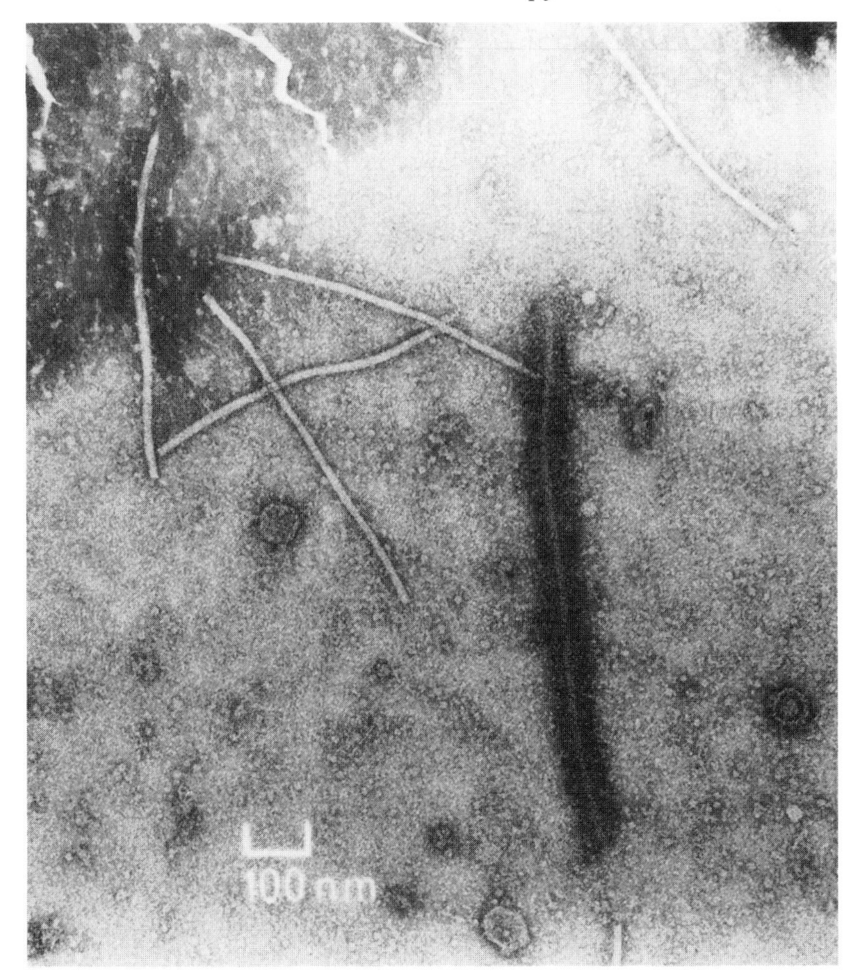

Fig. 6.14. Specifically labelled (decorated) dasheen mosaic virus in a mixture with unlabelled potato virux X.

of specific antiserum. It is most frequently used for the specific confirmation of the serological identity of unknown viruses (Fig. 6.14) or separating morphologically identical viruses in mixtures (Fig. 6.15). However, the technique can be used for antiserum titration. The titre, i.e. the highest dilution giving reliably detectable decoration, is usually one or two twofold steps higher than the gel diffusion or precipitin titre.

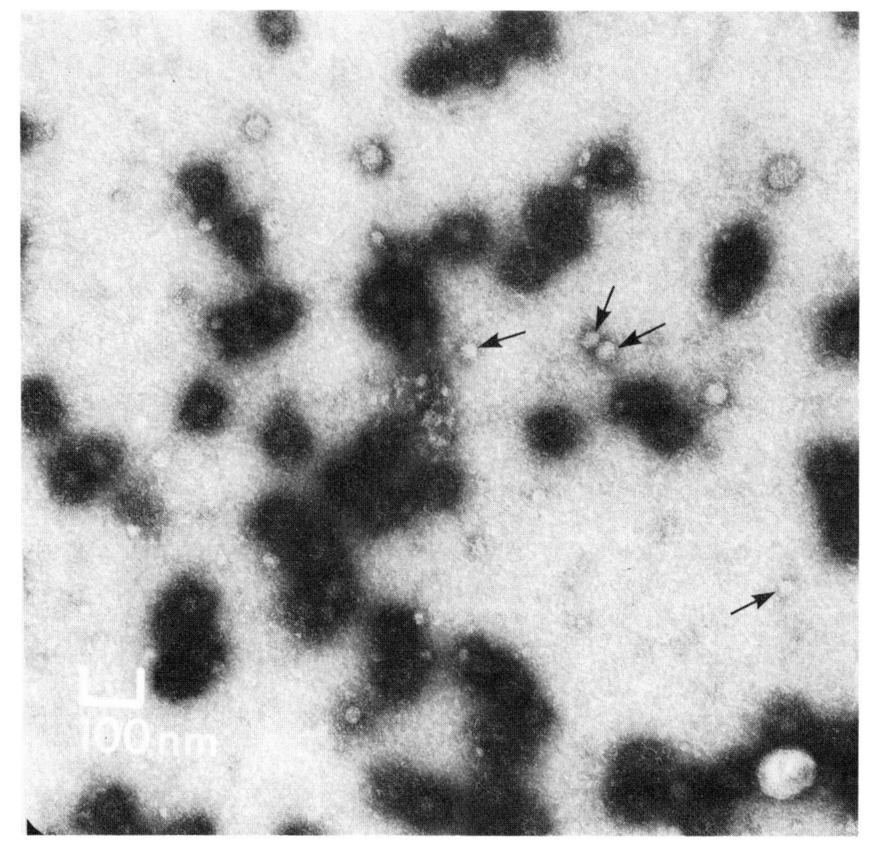

Fig. 6.15. Morphologically identical viruses. Poinsettia mosaic virus, surrounded by antibody decoration and Poinsettia cryptic virus undecorated (arrowed).

Materials

As for ISEM (p. 147), except crude antisera or purified gamma globulin fractions may be used. For diagnostic purposes dilutions of between 1/10 and 1/100 usually give a good strong coating layer; however, for antiserum titre determinations a dilution series is obviously required. The presence of host plant antibodies in the serum used does not interfere with the test since individual virus particles are observed directly. 'Dirty' antisera may therefore be used without the need for absorption of unwanted host plant antibody.

Procedure

1 Prepare the virus extract as for ISEM.

2 Touch freshly carbon-coated grids to a drop of the extract for a few seconds, rinse with 20 drops phosphate buffer then drain (but do not dry).
3 Float grids (with virus) support film downwards on diluted antiserum drops in a waxed Petri dish (or on Parafilm).
4 Incubate for 15 min at room temperature in the Petri dish ensuring that drops do not evaporate.
5 Wash the grid with 30 drops of water from a Pasteur pipette.
6 Stain by rinsing the grids with 6 drops of the appropriate stain.
7 Drain and allow to dry.

The 'decoration' technique works well with quick dip or squash preparations as described above. However, it may be used to even greater effect with virus trapped on the grid using the ISEM technique. Only a small number of virus particles are required for decoration, since antibody coating should be readily detected. When the two techniques are combined virus detection and characterization are achieved with great sensitivity.

Combined ISEM and decoration
1 Proceed through steps (**1**)–(**9**) of the ISEM procedure.
2 Rinse the grids with 20 drops of phosphate buffer using a Pasteur pipette.
3 Follow steps (**3**)–(**7**) of the decoration procedure.

As with ISEM, times and temperatures of incubation to achieve optimum decoration vary according to need. Incubation times from 1 min to 24 hours have been used and temperatures from 4 to 37°C. In general, the comments for ISEM apply, and time and temperature components may be adjusted as convenient.

Clumping (antiserum virus mixtures) (Fig. 6.16)

In this procedure the aim is to link virus particles with antibody bridges, and detect the aggregates in the electron microscope. Virus antibody aggregates should not be too large and there should be sufficient excess antibody to coat the virus particles (Fig. 6.17). This method is used for the sensitive detection and identification of virus particles.

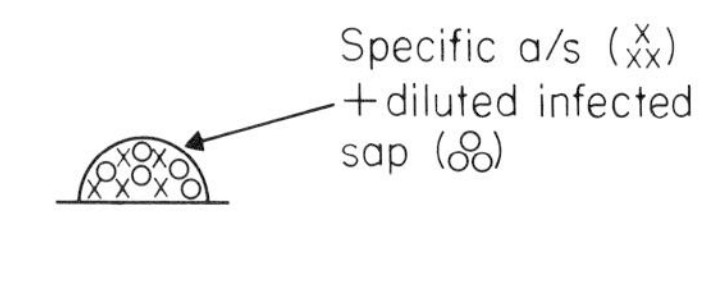

INCUBATE

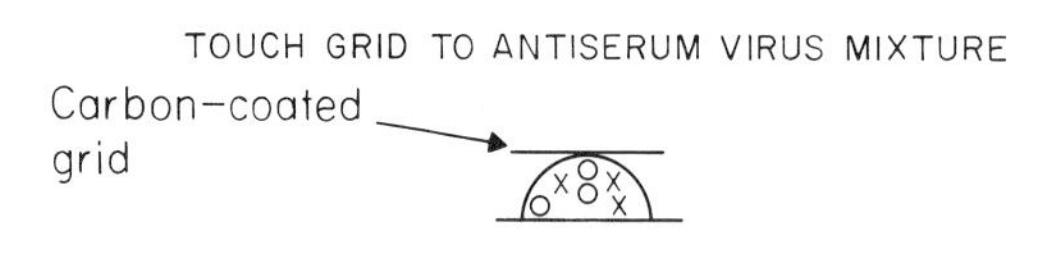

ADD NEGATIVE STAIN

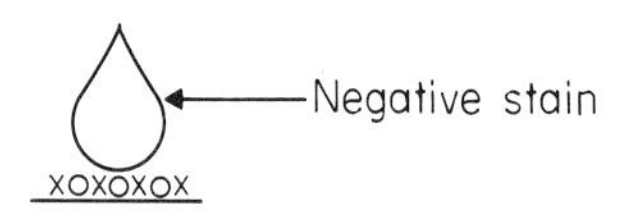

EXAMINE UNDER ELECTRON MICROSCOPE

Fig. 6.16. Clumping procedure.

Materials

As for ISEM (p. 147), except the following points.

1 Antisera: should be diluted in phosphate buffer, but not less than 1/16, and dilutions should preferably be made freshly each time they are used (see below).

2 Virus: should be extracted as for ISEM but should then be diluted in phosphate buffer until there is less than one rod-shaped filamentous particle per field of view in the electron microscope at a magnification of 20,000, or less than five spherical virus particles (conventional preparation).

Procedure

1 Pipette 40 μl of antiserum onto squares on a waxed Petri dish (or onto Parafilm).

2 Add 20 μl of diluted virus preparation to each antiserum drop.

3 Mix drops by carefully drawing up and expelling liquid from the pipette, leaving drop with a 'tight' meniscus.

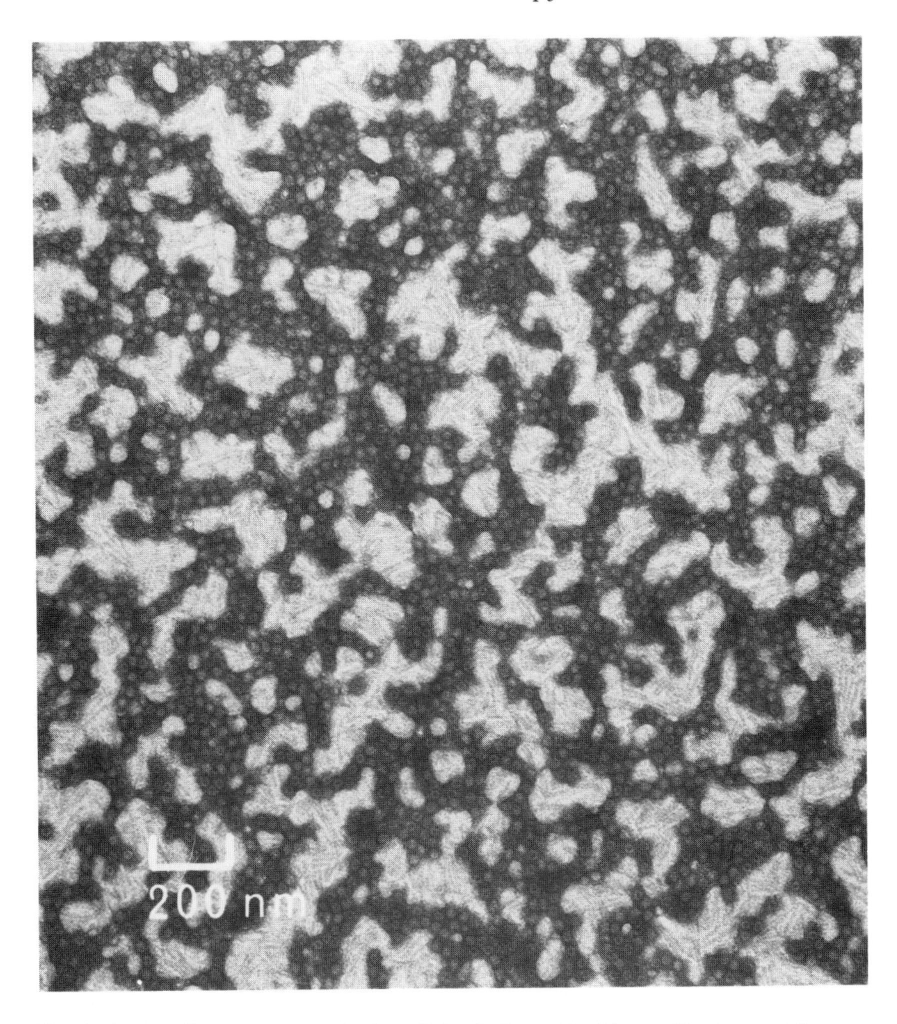

Fig. 6.17. Broad bean true mosaic virus. Spherical particles 'clumped' using specific antiserum. Concentration of virus particles is rather high for best reaction.

4 Cover Petri dish with lid to avoid evaporation and incubate either at 37°C for 0·5–3 hours at 4°C for 3–36 hours. The former gives rapid results for heat-stable viruses, the latter more even coating of antibodies on particles in aggregates.

5 After incubation, touch surface of drop with a carbon-filmed grid (coated side down).

6 Remove immediately and wash with 10 drops of stain (2% sodium phosphotungstate pH 6·8).

As with the other IEM techniques many variations are described. Shorter incubation periods at the higher temperature (or room temperature as a practical compromise) are often used to speed up the test. The dilutions of antiserum are less than with the decoration technique since both specific aggregation and decoration by antibody are required. If the antiserum concentration is too high however, decoration will occur but no clumping. Note that if purified or partially purified preparations are used, clumping or aggregation of particles may already have occurred despite IEM.

Uses of IEM

IEM techniques have found widest use in plant virus diagnosis for those viruses normally occurring in the plant in concentrations too low to be detected by conventional electron microscopy, e.g. the luteoviruses. Additionally, the use of antisera for specific confirmation of serological identity of viruses is of great value, being rapid and economical. However, as stated above the techniques have all the attributes of conventional serological tests and are at least as sensitive as most. IEM can be used to detect plant viruses within their vector [7] or for specific investigation of virus within precipitin bands in gel diffusion tests. Modifications of the ISEM and decoration procedures—to allow partial preparation, posting to another laboratory for completion at a later time, and return for examination—are under investigation. Such interrupted processing would render the sophistication of IEM detection available to more primitive or less well equipped laboratories.

As with conventional electron microscopy quantitative assay using IEM is possible [2].

In a number of investigations, mixed sera have been used in immunosorbent electron microscope procedures to detect more than one virus type in host sap. Usually the mixtures are of two, or at most three, different sera in sufficient buffer so that each is diluted to the optimum degree. Unless the viruses to be detected can be clearly differentiated morphologically, then the test may simply indicate virus presence or absence, not virus type. The sensitivity of such procedures using mixed sera has not been clearly described, but should suffer less from cumulative non-virus specific antiserum components than conventional serological tests, since virus particles are visualized.

References

1. Bradley, D.E. (1965) The preparation of specimen support films. In *Techniques for Electron Microscopy* (ed. Kay, D.) pp. 58–74. Blackwell Scientific Publications, Oxford.
2. Derrick, K.S. (1973) Quantitative assay for plant viruses using serologically specific electron microscopy. *Virology,* **56,** 652–653.
3. Hitchborn, J.H. & Hills, G.J. (1965) The use of negative staining in the electron microscope examination of plant viruses in crude extracts. *Virology,* **27,** 528–540.
4. Horne, R.W. (1965) Negative staining methods. In *Techniques for Electron Microscopy* (ed. Kay, D.) pp. 328–355. Blackwell Scientific Publications, Oxford.
5. Milne, R.G. (1972) Electron microscopy of viruses. In *Principles and Techniques in Plant Virology* (eds Kado, C.I. & Agrawal, H.O.) pp. 76–126. Van Nostrand Reinhold Co., New York.
6. Roberts, I.M. (1980) A method for providing comparative counts of small particles in electron microscopy. *Journal of Microscopy,* **118,** 41–245.
7. Roberts, I.M. & Harrison, B.D. (1979) Detection of potato leafroll and potato mop-top viruses by immunosorbent electron microscopy. *Annals of Applied Biology,* **93,** 289–297.
8. Shukla, D.D. & Gough, K.H. (1979) The use of protein A from *Staphylococcus aureus* in immune electron microscopy for detecting plant virus particles. *Journal of General Virology,* **45,** 533–536.

Appendix

Negative stains: abbreviated recipes

Phosphotungstic acid (PTA)

pH range 5–8.
Dodeca-tungstophosphoric acid dissolved in distilled H_2O to give 1 or 2% w/v solution.
Amend pH by dropwise addition of N NaOH.

Ammonium molybdate

pH range 3–9.
Finely ground solid dissolved in distilled H_2O to 2% w/v solution.
Amend pH by dropwise addition of N NaOH.

Uranyl acetate (UA)

pH range 3–4.
Solid dissolved in distilled H_2O to 1–2% w/v solution.
No pH amendment. Store cool and dark.

Uranyl formate

pH 3.
Solid dissolved in distilled H_2O to 1–2% w/v solution.
Store cool and dark. Prepare fresh regularly.

Methylamine tungstate

pH range 5–8.
2% w/v solution in distilled H_2O.
Amend pH by dropwise addition of N NaOH.

General use

Sørensens phosphate buffer: 0·066 M

Prepare stock solutions
A. 1/15 molar monopotassium phosphate (9·08 g KH_2PO_4 $litre^{-1}$ distilled water).
B. 1/15 molar disodium phosphate (11·88 g $Na_2HPO_4.2H_2O$ $litre^{-1}$ distilled water).
Mix A and B in the proportions

$$x \text{ ml A} + (100 - x) \text{ ml B},$$

where x is as follows for different pH values.

pH	x	pH	x
5·0	98·8	6·2	81·5
5·2	98·0	6·4	73·2
5·4	96·7	6·6	62·7
5·6	94·8	6·8	50·8
5·8	91·9	7·0	39·2
6·0	87·7		

Characteristics and preparation of support films [1]

Support films are of nitrocellulose (Collodion), polyvinyl, formaldehyde (Formvar), evaporated carbon, or combinations of these. Each material has particular characteristics which may or may not be desirable. Collodion and Formvar films are easily prepared without the need for vacuum coating equipment. The former is weak and may be unstable in the electron beam but provides a hydrophilic surface on which preparations spread well. The latter is stronger but more hydrophobic. Carbon films are strongest and most stable in the electron beam but may be brittle unless backed by Formvar or Collodion. They are thinner than plastic films and are therefore most suitable for high resolution work but may be hydrophobic. All kinds of support films become hydrophobic with time and are best used relatively fresh. However, coated grids may be kept for several weeks without undue deterioration if protected in a desiccator.

Preparation of support films

Whether preparing carbon, Collodion or Formvar films the process for actually applying the film to the grids is the same. In all three cases it is usual to produce a section of the appropriate film floating on the surface of dust-free water in a shallow container. The method for coating grids is as follows.

Materials

1 Shallow vessel (large Petri dish) or specially made settlement dish (Fig. 6.18).

2 Brass gauze (25×60 mm): must be kept grease free by washing in detergent.

3 Distilled water.

4 New clean copper grids.

5 Watchmaker's forceps.

Procedure

1 The gauze (with one corner bent up for ease of handling) is lowered to the bottom of the water-filled vessel and moved to one side.

2 Using fine forceps, copper grids are loaded onto the gauze until it is covered. Grids should not touch and should always be loaded with either their dull or shiny side uppermost: adopt a standard convention.

3 Float off support film as described below.

4 Raise gauze loaded with grids slowly upwards underneath support film catching it correctly orientated to cover the grids (Fig. 6.18).

OR allow the water to drain slowly from the specially

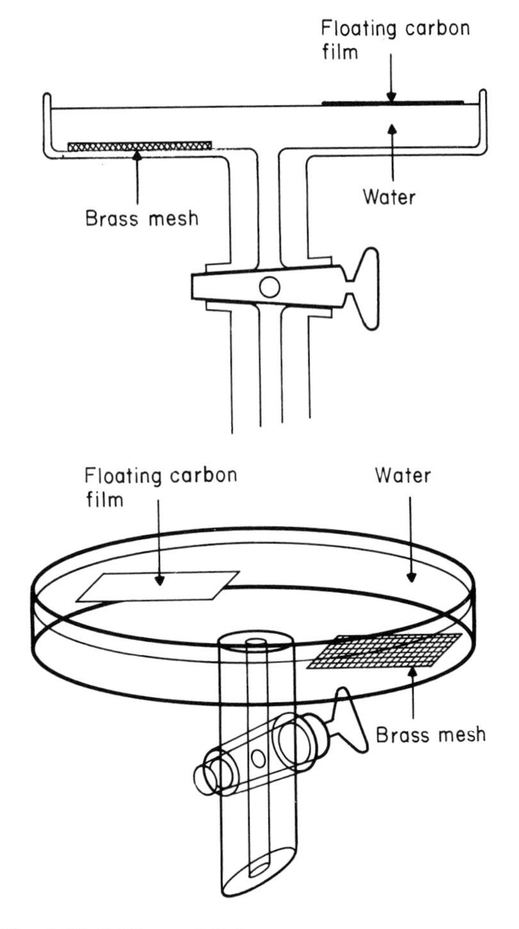

Fig. 6.18. Settlement dish.

manufactured vessel, steering the floating film so that it is lowered onto the grids on the gauze.

Support films

Formvar

1 This is prepared using the commercially available material in a solution of 0·2% w/v in chloroform or 1,2-dichloroethane (this should be stored in a spark-proof refrigerator).

2 A glass microscope slide (new unscratched) is cleaned using a chamois leather (not too clean or film may stick too well).

3 Using forceps dip the slide in the solution, withdraw it vertically and allow to dry.
4 Scrape the edges of the slide to cut the film, using the back of a scalpel.
5 Carefully, and slowly, lower the slide at an angle of about 30° into glass distilled water which should have a dust-free surface. The film should separate from the slide and float on the surface of the water (see Fig. 6.19).

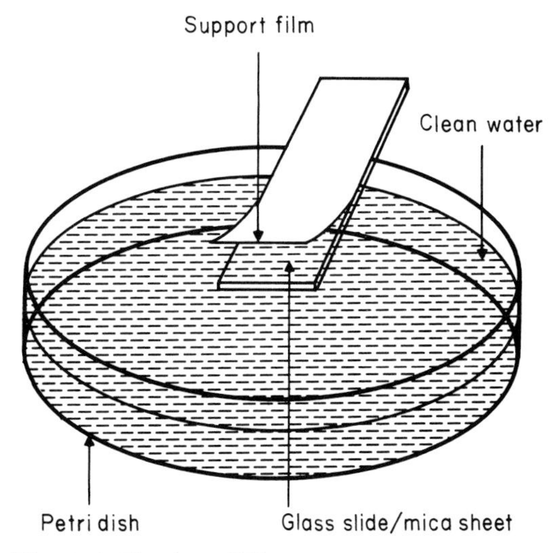

Fig. 6.19. Floating-off film.

6 Mount the film as described above, or before floating the film off the slide, cut it into grid-sized squares by scoring the slide carefully with a mounted needle. On floating off, the squares of film will separate and can be caught by individual grids held in forceps

Collodion

1 Prepare a solution of 2% Collodion in amyl acetate.
2 Allow a single drop of solution to fall onto the surface of clean glass-distilled water in a suitable container. The solution spreads leaving a film as the solvent evaporates.
3 Skim off the first film and discard it, this serves to clean the water surface.
4 Cast a second film as in (**2**) above.
5 Mount as described above.

Carbon

Preparation of carbon films requires the use of a vacuum coating unit (see Fig. 6.20). Within the evacuated bell jar of the coating unit an arc is struck by passing an electrical current between two pieces of carbon. Carbon vapour is formed which condenses on a suitable flat surface below.

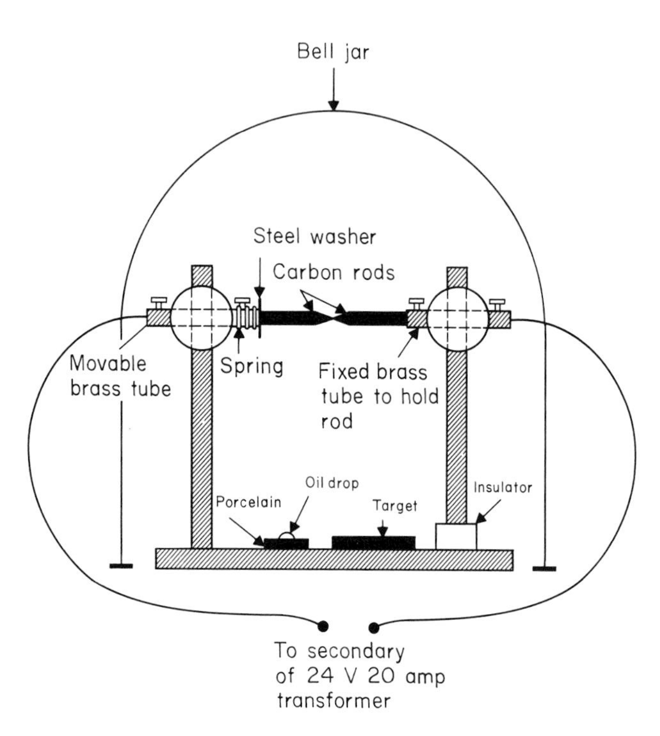

Fig. 6.20. Vacuum coating unit arrangement for carbon film preparation.

1 Prepare carbon rods by sharpening the ends, either
(a) one sharpened to a point and the other to a 45° face, or,
(b) both 'turned' so that their diameters are reduced and only the ends are pointed (see diagram).

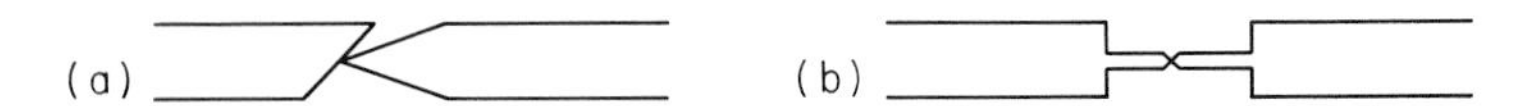

2 Fix the carbon rods, points just touching, in the spring loaded holder.

3 Place either a cleaned microscope slide, or a piece of freshly cleaned mica onto the floor of the vacuum chamber 15–20 cm directly below the carbon rods.
4 Beside the slide or mica, place a small piece of white porcelain onto which has been placed a small drop of vacuum oil.
5 Replace the bell jar and evacuate, preferably until the vacuum is better than 10^{-4} torr.
6 Strike an arc between the carbon rods (50A for about half a second) until the porcelain is straw coloured compared with the area protected by the oil drop. Only by trial and error will the best compromise between film thickness and stability be achieved.
7 Mount the film on grids as described above. For mica, before floating the carbon film off, trim the edges with scissors to free the film.

Suppliers of equipment

Alan Agar, Agar Aids, 66A Cambridge Road, Stanstead, Essex. Supply:

forceps;
settlement dish for film preparation;
support grids;
mica;
grid boxes;
carbon rod;
carbod rod grinder;
parafilm;
many other miscellaneous items.

BCL, Bell Lane, Lewes, E. Sussex. Supply:

small centrifuge tubes for IEM.

Emscope Labs, Kingsnorth Industrial Estate, Wotton Road, Ashford, Kent. Supply:

methylamine tungstate stain and miscellaneous items.

Taab Labs, 40 Grovelands Road, Reading, Berks. Supply:

'O' rings for forceps and miscellaneous items.

Index